# METHODS IN MOLECULAR BIOLOGY

*Series Editor*
John M. Walker
School of Life and Medical Sciences
University of Hertfordshire
Hatfield, Hertfordshire, AL10 9AB, UK

For further volumes:
http://www.springer.com/series/7651

# Microbial-Based Biopesticides

## Methods and Protocols

Edited by

**Travis R. Glare and Maria E. Moran-Diez**

*Bio-Protection Research Centre, Lincoln University, Lincoln, New Zealand*

Humana Press

*Editors*
Travis R. Glare
Bio-Protection Research Centre
Lincoln University
Lincoln, New Zealand

Maria E. Moran-Diez
Bio-Protection Research Centre
Lincoln University
Lincoln, New Zealand

ISSN 1064-3745 ISSN 1940-6029 (electronic)
Methods in Molecular Biology
ISBN 978-1-4939-6365-2 ISBN 978-1-4939-6367-6 (eBook)
DOI 10.1007/978-1-4939-6367-6

Library of Congress Control Number: 2016946365



Printed on acid-free paper

This Humana Press imprint is published by Springer Nature
The registered company is Springer Science+Business Media LLC New York

# Preface

The development of biopesticides based on microbes is an area of growing interest worldwide. Harnessing the power of naturally occurring antagonists of pests and diseases has always been an attractive proposal. We are surrounded by many instances of natural enemies keeping potential pests in check, so the idea of using natural enemies to reduce the pest issues, due in part to monoculture and intensive farming, has appeal. However, as has been repeatedly realised, a lot of research is needed to make this a reality for any specific pest.

The Methods in Molecular Biology (MiMB) series has been useful to many researchers, as few articles describe methods in sufficient detail to be able to reproduce without many learning errors. This can make learning a new techniques a frustrating and even costly business. Books which focus on the details of specific methods are much sought-after by researchers. This volume in the MiMB series is possibly pushing the envelope of what constituents molecular methods as many of the techniques are not all molecular based, but our aim is to provide methods of particular interest to those developing biopesticides based on live organisms. The area of biopesticide research and development is complex, ranging from selecting the right microbe to applying to the pest population; it requires cross-discipline science and industry cooperation.

A positive for biopesticide researchers is that there is a push to develop more sustainable pest control in most countries, with microbial-based pesticides an obvious choice. One aim of this book is to assist, in a small way, the wave of new developments of biopesticides, in the hope we can make the world a safer and healthier place.

We would like to thank all the contributors, for putting together their high quality and easy-to-follow protocols and share their knowledge with the scientific community. We also want to acknowledge John M. Walker and co-workers at Springer for the opportunity to broaden and gain substantial experience by assembling this collection of articles.

*Lincoln, New Zealand* — *Travis R. Glare*
*Lincoln, New Zealand* — *Maria E. Moran-Diez*

# Contents

# Contributors

HOSSEIN ALIZADEH • *Bio-Protection Research Centre, Lincoln University, Lincoln, New Zealand*
MARK BRAITHWAITE • *Bio-Protection Research Centre, Lincoln University, Lincoln, New Zealand*
CRAIG R. BUNT • *Department of Agriculture Sciences, Lincoln University, Lincoln, New Zealand*
LESLIE C.L. CHAN • *Patheon Biologics Australia Pty Ltd, Woolloongabba, QLD, Australia*
PETER CHEONG • *Bio-Protection Research Centre, Lincoln University, Lincoln, New Zealand; Biotelliga Limited, Pukekohe, New Zealand*
ANNABEL CLOUSTON • *Bio-Protection Research Centre, Lincoln University, Lincoln, New Zealand*
NICHOLAS CUMMINGS • *Bio-Protection Research Centre, Lincoln University, Lincoln, New Zealand*
ERNESTO FAVELA-TORRES • *Departamento de Biotecnología, Universidad Autónoma Metropolitana-Iztapalapa, Mexico City, Mexico*
INMACULADA GARRIDO-JURADO • *Department of Agricultural and Forest Science and Resources, University of Córdoba (UCO), Campus de Rabanales, CeiA3, Córdoba, Spain*
TRAVIS R. GLARE • *Bio-Protection Research Centre, Lincoln University, Lincoln, New Zealand*
ROMA L. GWYNN • *Rationale, Duns, Scotland*
STEPHEN R. HAINES • *AgResearch, Lincoln Research Centre, Christchurch, New Zealand*
JOHN HAMPTON • *Bio-Protection Research Centre, Lincoln University, Lincoln, New Zealand*
EMILY HICKS • *Bio-Protection Research Centre, Lincoln University, Lincoln, New Zealand*
ROBERT HILL • *Bio-Protection Research Centre, Lincoln University, Lincoln, New Zealand*
MARK R.H. HURST • *Innovative Farm Systems, AgResearch, Lincoln Research Centre, Lincoln, New Zealand*
MARK A. JACKSON • *United States Department of Agriculture, Agriculture Research Service, Crop Bioprotection Research Unit, National Center for Agricultural Utilization Research, Peoria, IL, USA*
SANDRA A. JONES • *Innovative Farm Systems, AgResearch, Lincoln Research Centre, Lincoln, New Zealand*
DIWAKAR KANDULA • *Bio-Protection Research Centre, Lincoln University, Lincoln, New Zealand*
JANAKI KANDULA • *Bio-Protection Research Centre, Lincoln University, Lincoln, New Zealand*
BLANCA B. LANDA • *Institute for Sustainable Agriculture (IAS), Spanish National Research Council (CSIC), Campus de Excelencia en Agroalimentación, CeiA3, Córdoba, Spain*
JIN-HUA LI • *Bio-Protection Research Centre, Lincoln University, Lincoln, New Zealand*
OCTAVIO LOERA-CORRAL • *Departamento de Biotecnología, Universidad Autónoma Metropolitana-Iztapalapa, Mexico City, Mexico*

LEILA MATINDOOST • *Australian Institute for Bioengineering and Nanotechnology, The University of Queensland, Brisbane, QLD, Australia*
AIMEE C. MCKINNON • *Bio-Protection Research Centre, Lincoln University, Lincoln, Christchurch, New Zealand*
ARTEMIO MENDOZA-MENDOZA • *Bio-Protection Research Centre, Lincoln University, Lincoln, New Zealand*
ROBERTO MONTESINOS-MATIAS • *Centro Nacional de Referencia de Control Biológico, Tecomán, Colima, Mexico*
JUAN A. MORALES-RAMOS • *Biological Control of Pests Research Unit, USDA-ARS NBCL, Stoneville, MS, USA*
MARÍA E. MORÁN-DIEZ • *Bio-Protection Research Centre, Lincoln University, Lincoln, New Zealand*
LIZZY A. MWAMBURI • *Department of Biological Sciences, University of Eldoret, El Doret, Kenya*
MARIA FERNANDA NIETO-JACOBO • *Bio-Protection Research Centre, Lincoln University, Lincoln, New Zealand*
ANGELA R. PAYNE • *United States Department of Agriculture, Agriculture Research Service, Crop Bioprotection Research Unit, National Center for Agricultural Utilization Research, Peoria, IL, USA*
ARNE PETERS • *e-nema GmbH, Schwentiental, Germany*
JAVIER PORCAYO-LOZA • *Departamento de Biotecnología, Universidad Autónoma Metropolitana-Iztapalapa, Mexico City, Mexico*
SALLY PRICE • *Department of Agriculture Sciences, Lincoln University, Lincoln, New Zealand*
CHARLOTTE PUSHPARAJAN • *Laboratory for Evolution and Development, Department of Biochemistry, University of Otago, Dunedin, New Zealand*
ENRIQUE QUESADA-MORAGA • *Department of Agricultural and Forest Science and Resources, University of Córdoba (UCO), Campus de Rabanales, CeiA3, Córdoba, Spain*
STEVEN REID • *School of Chemistry & Molecular Biosciences, The University of Queensland, Brisbane, Qld., Australia*
M. GUADALUPE ROJAS • *Biological Control of Pests Research Unit, USDA-ARS NBCL, Stoneville, MS, USA*
MICHAEL ROSTAS • *Bio-Protection Research Centre, Lincoln University, Lincoln, Christchurch, New Zealand*
DAVID I. SHAPIRO-ILAN • *United States Department of Agriculture, Agriculture Research Service, Southeastern Fruit and Tree Nut Laboratory, United States Department of Agriculture, Byron, GA, USA*
SCOTT STELTING • *USDA, Wildlife Services, National Wildlife Research Center, Fort Collins, CO, USA*
ALISON STEWART • *Bio-Protection Research Centre, Lincoln University, Lincoln, New Zealand; Marrone Bio Innovations, Davis, CA, USA*
JOHANNA STEYAERT • *Bio-Protection Research Centre, Lincoln University, Lincoln, New Zealand*
HERMANN STRASSER • *Institute of Microbiology, University of Innsbruck, Innsbruck, Austria*
JUDITH TAIBON • *Institute of Pharmacy, Department of Pharmacognosy, CCB – Centrum of Chemistry and Biomedicine, University of Innsbruck, Innsbruck, Austria; Institute of Microbiology, University of Innsbruck, Innsbruck, Austria*
GABRIEL VISNOVSKY • *Chemical and Process Engineering Department, University of Canterbury, Canterbury, New Zealand*
JESSICA YARDLEY • *Bio-Protection Research Centre, Lincoln University, Lincoln, New Zealand*

# Chapter 1

## What are Microbial-based Biopesticides?

María E. Morán-Diez and Travis R. Glare

### Abstract

From the ancient civilizations, agriculture has played an important socioeconomic role in the development of our current human society. Modern farming methods developed during the Green Revolution increased the production of food worldwide. Keeping a sustainable production of food supplies will impact the long-term survival of the species as well as the natural resources, so it is important to work on the establishment of new farming methodologies, such as the use of biopesticides that allow the balance between production and preservation of the environment. This introductory chapter will guide the reader through the content of this book providing an overall view of what will be discussed in each chapter.

**Key words** Biopesticides, Biological control, Integrated pest management (IPM), Green revolution (GR)

Have you ever found yourself looking for a rigorous, detailed protocol attempt at a new technique with no success? Felt the frustration of trying to interpret journal article methods sections and despaired at the incomplete descriptions? Many of us have experienced this at one time or another. Laboratory protocols are an extremely useful tool for those bench researchers who lack of enough experience in a specific field. Methods in Molecular Biology book series is a reference of basic procedures and offer an excellent opportunity to find complete, standardized protocols that are not commonly available in other published material. This edition will particularly benefit to those interested in biological control and includes step-by-step easy to follow methodologies in the field of microbial-based biopesticides. This book has been arranged in chapters, each of which has been written by recognized experts in the particular field of study based on their own trial-and-error experience. The aim of these chapters is to meet the needs of researchers and acts as a laboratory protocol handbook covering a wide range of techniques such as isolation, production, formulation, application, efficacy, and safety of biopesticides. This introductory chapter

Travis R. Glare and Maria E. Moran-Diez (eds.), *Microbial-Based Biopesticides: Methods and Protocols*, Methods in Molecular Biology, vol. 1477, DOI 10.1007/978-1-4939-6367-6_1, © Springer Science+Business Media New York 2016

attempts to give an overview of the past and present of microbial-based biopesticides and their implications in the development of a sustainable agriculture. Last chapter concludes with prospects and challenges for the future of biopesticides.

## 1 Integrated Pest Management in Sustainable Agriculture

The ever increasing world population is driving rising global food demand (FAO, 2014 "The Post-2015 Development Agenda and the Millennium Development Goals" report). However, food production is affected by an ever-growing number of threats, including social and political. Yet, biological factors are the single most important constraint to global food production. In fact, significant economic losses due to pests and diseases are estimated as 40% of total annual yield worldwide [1].

During the first half of the twentieth century, in an effort to combat pests, new programs of agriculture practices were launched in the framework of what it was called the "Green Revolution". Combination of policies of investment in breeding programs (mainly maize, rice, and wheat), distribution of inputs such as fertilizers, chemical pesticides, water regulation, and implementation of plantings in favorable agricultural regions, led to tripled crops yields in a short period of time [2]. It is undeniable the positive impact of this "Revolution" in terms of productivity, but despite initial success, a review of these agricultural policies [3] focused the main criticism on their ecological effects such as the use of monocultures and enormous amount of energy inputs. Even more important in terms of social opinion, the Green Revolution was criticized because of its dependence on the use of chemical pesticides. Subsequently, the ongoing increased use of agrochemicals has become a concern all over the world, often due to environmental and mammalian safety concerns, which has compelled international institutions to evaluate and undertake new policies to keep the chain of food demand and productivity rolling [3].

The concept of Integrated Pest Management (IPM) arose during the post Green Revolution period as a result of the emergence of problems associated with the application of agrochemicals, including broad spectrum activity that reduced naturally occurring ecosystem services. The search of alternative policies to minimize their use stimulated the application of microbe-based pesticides as an effective and environmental sensitive approach [1]. The international community agreed on the need for balancing productivity and environmental health as a key to achieve sustainability and preserve planet finite resources and ecosystems. Today, biopesticides as a part of IPM are considered essential in the implementation of agricultural sustainable programs. In fact, biopesticides have "come of age" [4].

## 2 What Are Microbial-Based Biopesticides?

The concept of microbial control, as the management of a pest by deliberate use of living organisms, may seem simple, but defining what the term microbial control covers in modern biological control is not. Today, there is a debate among authors and regulatory agencies about what "products" must be considered as biopesticides and, although there is still not unified definition for the term, biopesticides are generally used to describe the use of a wide range of formulated products based on biologically derived active agents to manage pests, diseases, and weeds [4]. According to US Environmental Protection Agency (EPA), these products can be shortlisted in three main categories: (1) microbial (bacteria, fungi, viruses, and protozoa) organisms and entomopathogenic nematodes as the active ingredient; (2) plant substances produced by genetically modified plants or plant-incorporated protectants (PIPs); and (3) naturally occurring substances like pheromones. In a recently published review [5], Glare illustrated the complexity of attempting to categorize these products, as previous authors had used approaches such as mode of action, targets, or origin of active ingredients. For instance, plant extracts are not accepted as biopesticides by some authors and regulators because their mode of action is based on toxic effects. Etymological arguments about term usage apart, a more important aim of defining biopesticides is to highlight the biological origin of the active agent as a control method and, therefore, their expected safety benefits compared to those of a chemical origin. It is also important to highlight that this debate does not only imply the assignment of a "label" but an obstacle in the process of registration of a new biopesticide, as is discussed in Chapter 16.

Biopesticides, if produced, formulated, and applied in appropriate ways, have been shown to have advantages over their synthetic competitors, especially lower nontarget toxicity including lack of mammalian toxicity.

In this book we have limited our consideration of biopesticides to the first category of the USA EPA that classifies pesticides based on live organisms such as bacteria, fungi, and nematodes (which are included under microbes for historic reasons in most jurisdictions). Metabolites from microbes are included under this definition in some cases where extraction processes are not specific, such as use of fermentation broths rather than just whole organisms. The first experiments involving biological control agents date back as far as the nineteenth century, when Agostino Bassi discovered that microbes cause disease using the fungus *Beauveria bassiana* on silkworms. Since then, an ever-growing number of studies and ideas for microbial-based biopesticides have been reported and products commercialized with some success. The most widely used biopesticide to this day is based on derivatives of the bacteria *Bacillus thuringiensis*

(Bt) isolated for the first time in the 1900s and first commercialized in France in 1938. Although there is no single and reliable comprehensive data source with information regarding the commercial current status of biopesticides (only private companies provides comprehensive reports but their availability is limited due to their cost), some regional databases keep up-to-date lists of products and provide physicochemical, toxicological, eco-toxicological, and other related data. One of these databases developed by the Agriculture & Environment Research Unit (AERU) in UK and used as a reference system across the EU is the Bio-Pesticides Database (BPDB) (http://sitem.herts.ac.uk/aeru/bpdb/) that includes around 600 biopesticides and their metabolites.

## 3 From the Field to the Bench and Back

Because of the implementation of IPM programs after the Green Revolution and the growing society awareness of potential environmental issues, the use of biopesticides has become more "popular" and many companies are attracted by the idea of developing eco-friendly products. Large agrochemical companies involved in the early period of biopesticides production usually failed because of the requirements for broad spectrum activity and high efficacy under all conditions. Today, this market is mainly occupied by small companies in local production and mostly relegated to organic agriculture [6]. Despite the many potential areas of use and the steadily growth in sales, estimated at over $400 million in 2012 [4], biopesticides have not yet achieved the anticipated market penetration and have done poorly in the world market for crop protection products where they only comprise 2–3 % of total sales. Reasons for the lack of competitiveness lie on the biological, technical, and regulatory challenges that new products face during the process of development and finally commercialization.

The process of development of a new microbe-based product is a path full of obstacles and can be a time- and money-consuming process. Many times strains which have been proven their efficacy when produced and applied by researchers in limited situations often perform poorly when tested on a larger scale. Therefore, the key for a more successful use of biopesticides must be focused on how to scale-up the use of biological control for today's modern-intensive agriculture. There are many issues to be addressed in this regards including cost of production, the quality of the inoculums, and the field efficacy of the organism. The Lubilosa program illustrates an excellent model of how to successfully address all these issues in order to counter a pest problem [7]. This program was established in 1989 in Africa and developed an oil-based formulation of the fungus *Metarhizium anisopliae* to control grasshoppers and locusts. The success of this program lies in the collaboration of researchers, industry, and farmers as a multidisciplinary team following a plausible step process of research, development, and marketing.

## 4 Methods in Biopesticide Research

The development of biopesticides advances from isolation and identification of beneficial organisms, through laboratory efficacy testing to production, formulation, and field application, before finally commercializing. Although it seems a linear process, it rarely follows this format, with many loops back to revalidate data or select new strains. Additional areas such as mode of action, response to ecological conditions, persistence, and safety have to be also considered. In this book, we have collected a number of useful methods across these areas of research. No single book could be exhaustive in coverage of techniques, but these chapters cover many areas from isolation to application (Fig. 1).

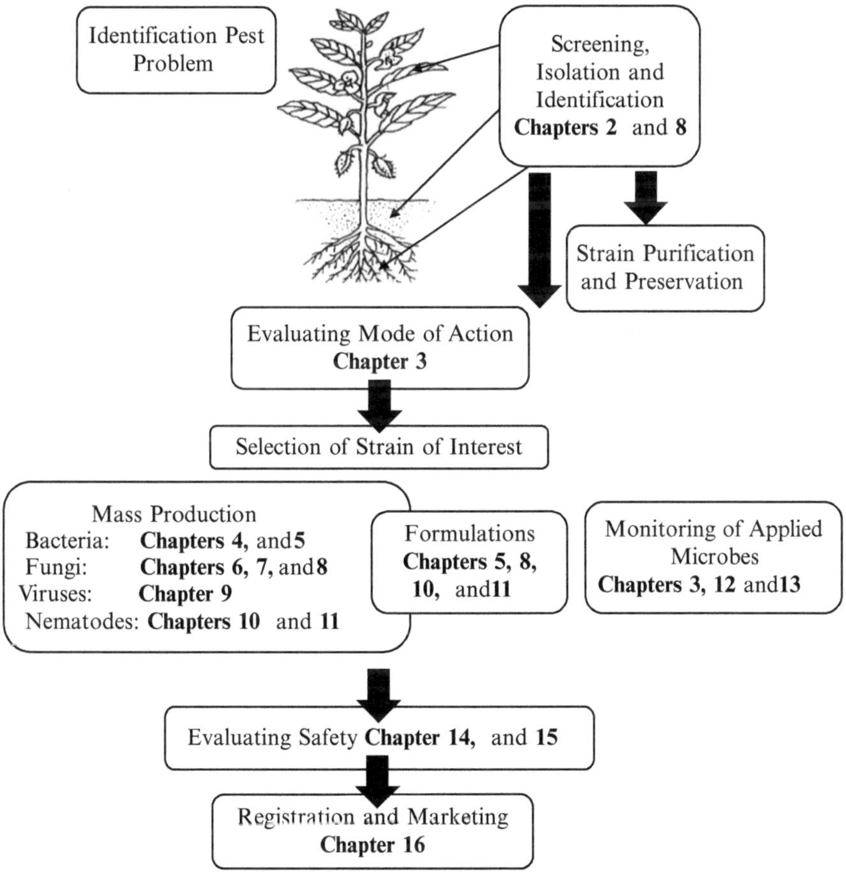


**Fig. 1** Schematic example of stage-based scheme from the initial phase of the pest problem identification to the final phase of marketing the product through methodologies described in this book

Once the pest problem is identified, researchers must acquire knowledge of the target pest, including pest biology and population dynamics in order to narrow the search of a control candidate. For instance, evidence of presence of beneficial organisms in a particular sampling area, such as dead insects or healthy plants, will provide a good starting point for screening and isolation [8]. Also, the mechanism of action of the biopesticide will be determined by the type of organism used (bacteria, fungi, or viruses) and therefore is important to focus on its taxonomy and biological characterization. The screening and isolation of a candidate among several hundreds of thousands of isolates can be a long and tedious process and requires a suitable mass-throughput method. Some techniques for selection include the use of phenotypic or genotypic markers, DNA arrays, or production of specific toxins or metabolites.

Techniques used to isolate potential beneficial organisms from soil and plant samples can be found in Chapters 2 and 8. Once the microorganism has been isolated it is important to evaluate their potential as bioactive agent before further research is done. Dual culture plate assays, production of volatile and nonvolatile compounds tests, detached necrotic leaf assays, or measurement of enhancing plant growth and systemic-induced response *in planta* are easy-to-perform procedures providing practical information on the potential behavior of an organism as a biological control agent (*see* Chapter 3).

Before mass production of the control organism it is critical to conduct subculturing procedures and preservation to ensure the genetic stability of the isolates and their viability. Most authors in this book include a section in their chapters where they describe the methodology used in their laboratories to culture, purify, and preserve the organism of interest.

The process of production does not only depend on the organism to be produced but also the final product to be used as bioactive agent. Hence, requirements for production of biological structures such as conidia, microsclerotia, or insecticidal viruses differ from those used in the purification of metabolic products secreted by these organisms. Yet, in general the production process may be categorized as based on liquid- or solid-phase fermentation, the latter of which in most cases requires a previous liquid phase to obtain the inoculum used in the solid production stage. Liquid culture fermentation would be the chosen method of production in cases of bacteria, nematodes, or fungi because it is a relatively low-cost process that can be scaled-up to large volumes. This process requires a stable inoculum and monitoring of parameters such as temperature, pH, dissolving oxygen, and nutritional component factors to ensure a quality yield of products [9]. Jackson and Payne describe in Chapter 7, a liquid culture fermentation technique used to produce stable microsclerotial granules.

This methodology is based on the capacity of some fungi to produce sclerotia, a survival structure that under suitable environmental conditions, strongly dependent on nutritional factors, produce conidia to infest insects or plant pathogens. The authors describe how these factors can be manipulated using liquid-stage cultures to induce the differentiation of sclerotia in fungi that have not been reported to produce sclerotia in nature, such as the biocontrol agents *Metarhizium* and *Trichoderma*. Another example of biological structures used as biopesticides and produced by liquid fermentation is described in Chapter 9. The authors present an exhaustive methodology addressed to optimize and establish a small-scale production of insect cells for production of baculovirus that later may be easily scaled up.

As mentioned before, final products of liquid fermentations include not only biological structures but also metabolites. Chapter 4 focuses on the production of an insect-active toxin complex secreted by the bacteria *Y. entomophaga* and its purification through size exclusion chromatography (SEC).

In contrast to the submerged state, gradients such as temperature, pH, or dissolving oxygen are difficult to control under water depravation in solid-stage fermentation, which makes this technology an unsuitable choice for industry purposes [10]. However, it has some advantages for a small-scale production: is the best method of obtaining fungal spores by aerial hyphae, such is the case of *Beauveria* species, and also reduces the possibilities of contamination by bacteria and yeast due to the low availability of water. Because of the lower levels of moisture solid-state production is rarely used for mass bacteria production. The methodology developed by Loera-Corral et al. (*see* Chapter 6) compares two systems: plastic bags, using rice as substrate for conidia development, and tubular bioreactors as an easy-to-follow monitoring method. A similar concept based on the use of substrates such as barley, rice, or wheat bran is developed in Chapters 8 and 2 where authors describe a low-cost and easy methodology to produce conidia of *Beauveria bassiana* and *Trichoderma* species, respectively.

An in vivo approach, instead of an in vitro fermentation processes, is generally used for the propagation of some entomopathogenic nematodes. This biological control organism has an active metabolism and needs high humidity and oxygen to survive. It has been reported that the use of insect host for small applications is the best choice for nematodes production and carrier for application [11]. Shapiro-Ilan and his group detail in Chapter 11 a methodology based on these features where they use an automated procedure to improve production, formulation, and packing host infected insects. This chapter provides a detailed description of all steps to follow to successfully produce nematodes at small scale.

All these methodologies previously described require working with stable stock cultures that consistently produce the same performance as the original stock at small scale and achieve the highest yield possible with the lowest cost of culture medium. In doing it so, it is necessary that a suitable methodology for storage and preservation of the bioactive organism is used to increase the shelf-life. Formulation is, therefore, an essential step in the development of a viable biopesticide. Sometimes data available in this regards is limited due to patents and industry confidentiality what makes of this book a useful source of information.

Critical formulation requirements are determined by the organisms themselves and by their environments where they will be applied. It is for this reason that knowledge of the host–pathogen–biocontrol relation is crucial for the success of the biopesticide in the field. The aim of formulation is to stabilize the organism during production, distribution, and storage; ensure good performance and persistence at the target site; and be easy to handle and apply. This can be achieved by liquid or dry formulations. The first method is based on oil, water, polymers, or a combination and, because biopesticides are usually live organisms, requires adding an inert ingredient such as stabilizers or surfactants to improve stability and make them easier to handle and apply. Dry formulations are produced by adding binder, dispersant, or wetting agents [12]. A good practice is to store them in water vapor-proof containers to avoid their deterioration. There is not a magical recipe and each method is established by trial-and-error assay. Hence, Chapter 10 provides a useful guide of techniques addressed to obtain formulations of nematodes under in vitro conditions. The author describes methodologies based on the use of wettable powder and alginate beads. Some carriers, such as clay, kaolin, or peat, are commonly used in formulations to enhance shelf-life and efficacy of the final product (*see* Chapter 8). For an overview of the main features to consider when developing formulations based on coating techniques, refer to Chapter 5.

Application methodology is undoubtedly linked to the process of formulation because the way to apply the biopesticide determines its formulation and *vice versa*. In many cases, the application system is similar to chemical pesticides and similar equipment can be used to make it more economically efficient.

This book also contains two chapters about a growing area of interest, endophytes, which are fungi and bacteria that occur and develop within the plant. Glare et al. [4] include these organisms as a class of biopesticides as they grow asymptomatically within plant tissue and can promote plant defenses against insects and diseases. It is important to know the interaction between endophyte and plant to develop the right product and to do it so it is necessary to screen methodologies based on detection of these organisms that will help to determine their life style and therefore to make better formulations and applications. A protocol to detect *Beauveria bassiana* by quantitative PCR, which may also be applied to other fungi endophytes, is described in Chapter 12. A critical step to be considered in order to achieve accuracy detection of endophytes relies on the preparation of plant tissues *prior* to any detection analysis is performed. This is due to the epiphytic community found on host surfaces that may be positively identified as a false endophyte if samples have not been correctly surface sterilized. McKinnon provides in Chapter 13 a protocol based on the sterilization of plant surfaces to accurately identify fungal endophytes.

The regulatory authorities still have concerns regarding the use of biopesticides and data about the composition, toxicity, or degradation is required to be provided in order to register a new product. Some procedures for detection and quantification of fungal metabolites are described in Chapter 15. Chapter 14 details some protocols to measure the activity of some enzymes in *B. bassiana* although these procedures can be easily used to measure their activity in other fungi.

By this point of the development process, the science is somehow replaced by a more commercial and bureaucratic approach. Economic and regulatory considerations must be considered as important as the scientific process since they will influence the failure or success of commercialization. As we mentioned earlier, the registration process is one of the most time and money consuming. Chapter 16 provides an overview of the many issues biopesticides have to face to in order to be commercialized.

There is still scope for the improvement but more investment in research and development and collaboration among researchers, institutions, and growers is needed and only understanding of the usefulness of biopesticides as a part of a sustainable agriculture system will make possible a quantitative and qualitative jump forward to make microbial-based pesticides the option of choice.

Although the chapters are listed under specific topic for a better understanding of the process of developing a biopesticide, most authors, as pointed in this introductory chapter, provide not only a single method, such as a production technique, but also how to formulate or apply a formulation. This is due to production, formulation, and application phases are intimately interrelated and methodologies are correlated with each other. However to assist reading this book, we present the contents based on the main methodology described in each chapter (*see* Fig. 1 for a better understanding and visualization of links among topics and chapters).

## References


1. Chandler D, Bailey AS, Tatchell GM, Davidson G, Greaves J, Grant WP (2011) The development, regulation and use of biopesticides for integrated pest management. Philos Trans R Soc Lond B Biol Sci 366:1987–1998
2. Pingali PL (2012) Green revolution: impacts, limits, and the path ahead. Proc Natl Acad Sci U S A 109:12302–12308
3. Glaeser B (2010) The green revolution revisited: critique and alternative. Routledge Taylor & Francis Group, London
4. Glare T, Caradus J, Gelernter W, Jackson T, Keyhani N, Kohl J, Marrone P, Morin L, Stewart A (2012) Have biopesticides come of age? Trends Biotechnol 30: 250–258

5. Glare T (2015) Types of biopesticides. In: Nollet LML, Rathore HS (eds) Biopesticides handbook, 1st edn. Taylor & Francis Group, London, pp 7–24
6. Ravensberg WJ (2011) Roadmap to the successful development and commercialization of microbial pest control products for control of arthropods. Springer Science & Business Media, New York
7. Douthwaite B, Langewald J, Harris J (2001) Development and commercialization of the green muscle biopesticide. International Institute of Tropical Agriculture CABI Biosciences, UK
8. Montesinos E (2003) Development, registration and commercialization of microbial pesticides for plant protection. Int Microbiol 6:245–252
9. Fulekar MH (2010) Environmental biotechnology. CRC Press, Taylor & Francis Group, London
10. Holker U, Hofer M, Lenz J (2004) Biotechnological advantages of laboratory-scale solid-state fermentation with fungi. Appl Microbiol Biotechnol 64:175–186
11. Shapiro-Ilan DI, Han R, Qiu X (2014) Production of entomopathogenic nematodes. In: Morales-Ramos J, Rojas G, Shapiro-Ilan D (eds) Mass production of beneficial organisms: invertebrates and entomopathogens. Academic, Amsterdam, pp 321–356
12. Burges HD (1998) Formulation of microbial biopesticides: beneficial microorganisms, nematodes and seed treatments. Springer Science, New York

# Part I

## Screening, Isolation and Identification of Potential Biological Control Agents

# Chapter 2

# Isolation and Mass Production of *Trichoderma*

**Artemio Mendoza-Mendoza, Annabel Clouston, Jin-Hua Li, Maria Fernanda Nieto-Jacobo, Nicholas Cummings, Johanna Steyaert, and Robert Hill**

## Abstract

Members of the genus *Trichoderma* comprise the majority of commercial fungal biocontrol agents of plant diseases. As such, there is a wealth of information available on the analysis of their biocontrol potential and the mechanisms behind their superior abilities. This chapter aims to summarize the most common methods utilized within a *Trichoderma* biocontrol program for the isolation, identification, and mass propagation of individual strains.

**Key words** *Trichoderma*, Isolation, Identification, Conidia, Conidial propagation

## 1 Introduction

*Trichoderma* species are versatile, ubiquitous filamentous fungi which are found free living in soil, colonizing dead organic matter, and in beneficial endophytic associations with plant species. Collectively, *Trichoderma* species comprise the majority of commercially available fungal agents for the control of plant diseases (caused by fungi, bacteria, and nematodes). The majority of commercial preparations consist of conidia which are the asexual reproductive spores of this genus; therefore, the ability to reproduce well in culture is an essential feature of a successful biocontrol agent [1]. The first step in any successful *Trichoderma* biocontrol research program is the isolation of candidates and typically this is achieved by either bioprospecting suppressive soils for free-living strains or by extracting endophytic strains from the roots of healthy plants [2]. The second step is the bulk propagation of conidia for subsequent biocontrol trials. Effective biocontrol is achieved through a combination of mechanisms including mycoparasitism, competition for nutrients and/or space, antibiosis, and induction of systemic resistance [3–8]. Additionally, some *Trichoderma*

Travis R. Glare and Maria E. Moran-Diez (eds.), *Microbial-Based Biopesticides: Methods and Protocols*, Methods in Molecular Biology, vol. 1477, DOI 10.1007/978-1-4939-6367-6_2, © Springer Science+Business Media New York 2016

strains induce temperature and drought tolerance [5, 8], while other strains induce plant growth promotion and effectively act as biofertilizers [9, 10]. In this chapter, we describe diverse methods for the isolation and propagation of *Trichoderma* biocontrol candidates.

## 2 Materials

### 2.1 General Media

All media are autoclaved at 121 °C for 15 min, cooled to 50 °C, and poured into 90 mm Petri dishes, unless otherwise stated. Standard media (e.g., potato-dextrose agar (PDA) and broth (PDB); malt extract agar (MEA) and broth (MEB); Malt–Yeast extract broth (MYE) is prepared according to manufacturer's instructions.

1. *Trichoderma* selective medium (TSM): 10.0 g malt extract, 1.0 g yeast extract, 0.2 g Terraclor (quintozene) and 0.15 g Rose Bengal, add 900 mL deionized water (DW) and stir until dissolved. Add 15.0 g agar, and fill up to 1000 mL with DW. Bring to the boil in the microwave (6–7 min) and add 0.6 mL chloramphenicol stock solution (100 mg/mL) before autoclaving. Alternatively add 10 mL of filter-sterilized streptomycin/chlortetracycline solution after autoclaving. Plates should be stored at 4 °C.
2. Antibiotics stock solution for TSM: Weigh 2.5 g streptomycin sulfate and 0.5 g chlortetracycline HCL, add 100 mL of DW and stir until fully dissolved. Filter the solution to 0.2 μm and store aliquots at −20 °C.
3. MYE: Suspend 20 g of malt extract and 2 g of yeast extract in 1 L distilled water. Mix thoroughly to dissolve and sterilize by autoclaving.

### 2.2 Isolation

1. 1 % Virkon.
2. Universal bottles containing 9 mL SDW with 0.1 % Tween 80.
3. Wrist action shaker.
4. SDW with 0.1 % Tween 80.
5. PDA plates with 0.2 % Triton X100.

### 2.3 Mass Production of Trichoderma Inoculum on Brown Rice and Wheat Grain

1. Brown rice and wheat grain.
2. SDW with 0.1 % Tween 80.
3. ½ PDB or liquid MYE.
4. Cotton wrapped in cheesecloth (cotton bung).

## 3 Methods

### 3.1 *Trichoderma* Isolation from Plant Roots

1. Select healthy plants and using a spade or trowel dig up a small section of root material. Store in a labeled, moist, zip-lock bag. Keep the root sample moist by using a spray bottle to mist the root and the inside of the bag (*see* **Note 1**).
2. Record the plant genus and species (if possible), site location, and any other relevant details. Take a photo of the plant/site if possible.
3. Wash the root section under tap water to remove bulk soil. Cut approximately 30×1 cm pieces of root sections from the sample.
4. Soak 24 pieces of root in a deep Petri dish containing ~10 mL of 1 % Virkon (*see* **Note 2**) for 10 min to sterilize the surface. Soak the remaining 6 pieces in another deep Petri dish containing ~10 mL sterile distilled water (SDW) (for isolation of fungi from the root surface).
5. Rinse three times in SDW and shake off excess water, then briefly air dry. Using three TSM and one PDA plate, plate six pieces per plate. The six root pieces washed in SDW only are plated onto one TSM plate (Fig. 1).
6. Seal and incubate plates at room temperature (~20 °C) for 7 days, at which time colonies can be subcultured onto PDA to obtain pure cultures.

### 3.2 *Trichoderma* Isolation from Rhizosphere and Bulk Soil Samples

1. Select healthy plants and using a spade or trowel dig up a small section of root material. Using a small spatula, collect 15 g of rhizosphere soil around the roots (rhizosphere), or bulk soil (*see* **Note 3**). Store in a labeled, moist, zip-lock bag at 4–8 °C until ready for processing.
2. Record the plant genus and species (if possible), site location, and any other relevant details. Take a photo of the plant/site if possible.
3. Add 1 g soil to 9 mL 0.1% Tween 80 in universal bottles and place on a wrist action shaker for 10 min at maximum speed and then leave to stand for 10 min. Dilute 100, 1000, and 10,000-fold and plate 100 μL onto 3×TSM plates for each dilution.
4. Seal and incubate plates at room temperature (~20 °C) for 7 days, at which time colonies can be subcultured onto PDA to obtain pure cultures.

### 3.3 *Trichoderma* Conidia Production on Plates

1. Centrally inoculate PDA plates with either 5 μL of *Trichoderma* conidial suspension or an agar plug from the margin of an actively growing *Trichoderma* colony.

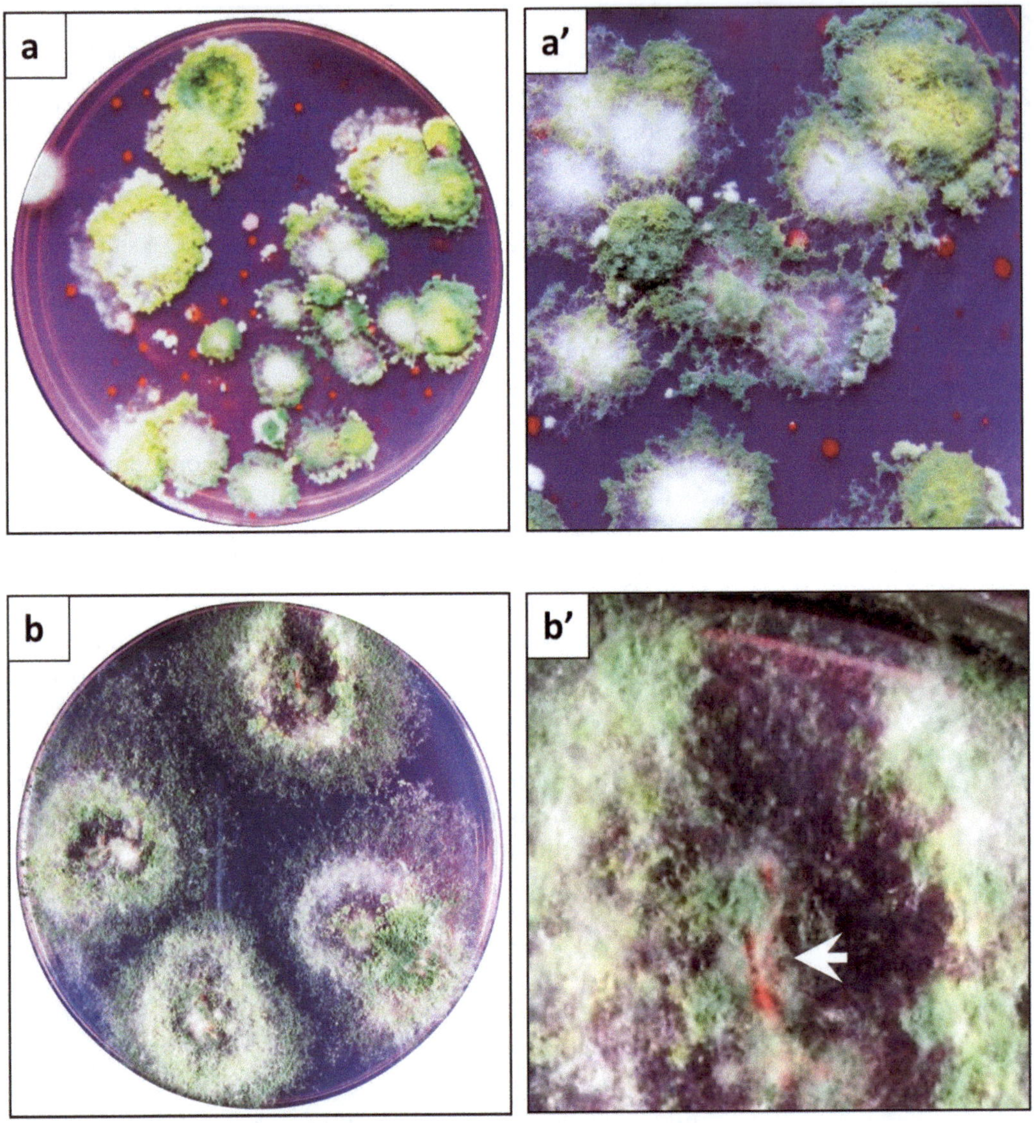


**Fig. 1** *Trichoderma* spp. isolated from soil (**a-a'**) and plants (**b-b'**). TSM was used for selecting *Trichoderma*. **a'** and **b'** are a close up to indicate the morphology of the *Trichoderma* colony (**a'**) and the plant material plated in the TSM *white arrow* (**b'**)

2. Seal the plates with millipore tape and incubate it at 25 °C for 7–10 day in a 12 h/12 h light/dark cycle to encourage conidiation. Alternatively maintain the plates in constant light for the same period and temperature.
3. Add 10 mL 0.1% Tween 80 and rub the colony with a sterile rod to liberate the conidia. Filter the conidial suspension through two layers of Miracloth (Merck Millipore) (*see* **Note 4**) inside a funnel placed into a sterile universal bottle.

4. For preservation of conidia, to 500 μL of conidia suspension add 500 μL of 50% glycerol into a cryotube, vortex the mixture and store at −80 °C.

#### 3.4 Strain Purification: Isolation of Colonies Derived from a Single Conidium

1. Harvest and filter conidia from a freshly sporulating culture as described in Subheading 3.3.
2. Dilute the conidial suspension $10^{-6}$-fold with SDW/Tween 80 and plate 250 μL onto the surface of a PDA + Triton X plate (*see* **Note 5**). Spread with a sterile rod.
3. Seal plates and incubate at 20 °C in the dark (*see* **Note 6**) for 2–5 days. Check plates every day under a stereo microscope and subculture a single germinating conidium onto fresh PDA.

#### 3.5 Mass Production of Trichoderma Inoculum on Brown Rice

1. Weigh 300 g of brown rice and place in a 30 × 66 cm autoclave bag, add 225 mL tap water and mix thoroughly.
2. Roll up the bag loosely, leaving enough space for evaporation of water when autoclaving, and tape up the edge of the bag.
3. Place three bags of rice in a metal autoclave tray, enclose in a large autoclave bag, and sterilize at 121 °C for 15 min.
4. Aliquot the autoclaved brown rice into sterile polypropylene food trays, adding approximately 100 g of rice per tray.
5. Prepare *Trichoderma* conidia as described in **steps 1–3** in Subheading 3.3, but instead adding 7.5 mL of ½ PDB or liquid MYE to detach conidia from the plates.
6. Pour the conidial suspension onto brown rice in a sterile polypropylene tray (one plate per tray) and mix.
7. Place each tray in a zip-lock plastic bag and seal, leaving a slight opening on one side.
8. Incubate at room temperature close to a window for exposure to natural lighting. Most *Trichoderma* isolates will conidiate around 5–7 days postinoculation.

#### 3.6 Mass Production of Trichoderma Inoculum on Wheat Grain

1. Add 200 g high quality wheat grain and 0.0625 g chloramphenicol to 1 L Erlenmeyer flask.
2. Make up to 500 mL with distilled water and heat on hot plate until boiling.
3. Remove and stand for 10 min.
4. Drain water through cheesecloth.
5. Wash grain three times with tap water until clear.
6. Drain well.
7. Cover flask with cotton bung wrapped and aluminum foil.
8. Autoclave for 15 min at 121 °C.
9. 24 h later, wash grain with tap water and drain well.

10. Dispense grain into 250 ml conical flasks.
11. Autoclave for 15 min at 121 °C.
12. Inoculate the wheat grain with five mycelial plugs from the margin of an actively growing *Trichoderma* colony and incubate at 25 °C for 5–7 days, shaking flasks vigorously every day to avoid clumping.
13. Dispense grain into sterile plastic containers and incubate until profuse conidiation occurs (Fig. 2a).

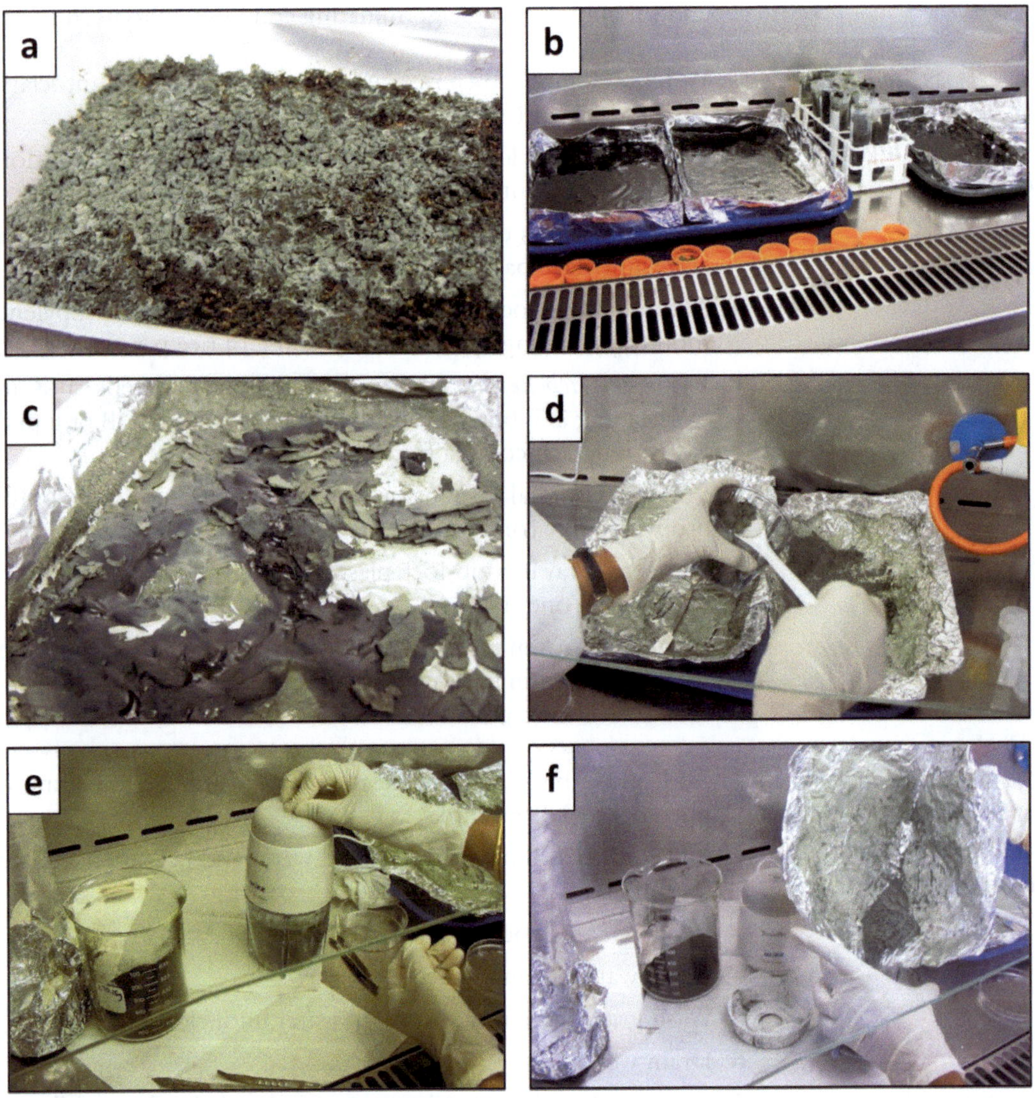


**Fig. 2** Mass production process of *Trichoderma* spp. spores. (**a**) Sporulation in sterile plastic containers. (**b**) Spore suspension on aluminum foil. (**c**) Spores dried overnight in a sterile hood. (**d-e**) collect and grinding spores. (**f**) Final fine spores powder

14. Collect the conidia using SDW, pass through sterile cheesecloth, centrifuge and place the conidial suspension on an aluminum foil for overnight drying in a laminar hood (Fig. 2b).
15. Dried spore flakes are taken in a grinder and ground to a fine powder (Fig. 2c–f).

## 4 Notes

1. Process the samples the same day or keep the samples in a fridge for no more than a day or two.
2. Alternatively to Virkon, use a 70% (v/v) ethanol solution for 5 min followed by incubation for 5 min with 5% sodium hypochlorite. Use a common household bleach and check the percent active ingredient (sodium hypochlorite)—typically 3–5%. Dilute with water to achieve the required sodium hypochlorite percentage.
3. Rhizosphere soil is defined as the soil within 1 cm of the root. Bulk soil is defined as soil not affected chemically, physically, or biologically by the root of the plant.
4. Filtering the conidial suspension through Miracloth removes hyphal fragments.
5. Triton X limits the diameter of fungal colonies.
6. Maintaining the plates sealed and in the dark helps prevent photoconidiation.

## Acknowledgements

The work of the *Trichoderma* team from Lincoln University has been supported by the Bio-Protection Research Centre, a Centre of Research Excellence and by the following New Zealand funding agencies/sources: Tertiary Education Commission; Foundation for Research Science and Technology; Ministry for Business, Innovation and Employment; Agricultural and Marketing Research and Development Trust; Marsden Research Fund; Lincoln University Research Fund; Lincoln-Massey Research Excellence program.

### References

1. Steyaert JM, Weld RJ, Mendoza-Mendoza A, Stewart A (2010) Reproduction without sex: conidiation in the filamentous fungus *Trichoderma*. Microbiology 156:2887–2900
2. Hill RA, Ambrose A, Sajali NA, Yatim M, Valdez RB, Agbayani F, Bungang J, Minchin R, Stewart A (2010) Bioprotection of *Acacia mangium* using *Trichoderma* in Malaysia. In: Zydenbos SM, Jackson TA (eds). Microbial products: exploiting microbial diversity for sustainable plant production. New Zealand Plant Protection Society Inc. Christchurch, New Zealand 51–56
3. Vinale F, Flematti G, Sivasithamparam K, Lorito M, Marra R, Skelton BW, Ghisalberti

EL (2009) Harzianic acid, an antifungal and plant growth promoting metabolite from *Trichoderma harzianum*. J Nat Prod 72: 2032–2035

4. Vinale F, Ghisalberti EL, Sivasithamparam K, Marra R, Ritieni A, Ferracane R, Woo S, Lorito M (2009) Factors affecting the production of *Trichoderma harzianum* secondary metabolites during the interaction with different plant pathogens. Lett Appl Microbiol 48:705–711
5. Druzhinina IS, Seidl-Seiboth V, Herrera-Estrella A, Horwitz BA, Kenerley CM, Monte E, Mukherjee PK, Zeilinger S, Grigoriev IV, Kubicek CP (2011) *Trichoderma*: the genomics of opportunistic success. Nat Rev Microbiol 9:749–759
6. Calistru C, McLean M, Berjak P (1997) In vitro studies on the potential for biological control of *Aspergillus flavus* and *Fusarium moniliforme* by *Trichoderma* species. A study of the production of extracellular metabolites by Trichoderma species. Mycopathologia 137:115–124
7. Mukherjee PK, Buensanteai N, Moran-Diez ME, Druzhinina IS, Kenerley CM (2012) Functional analysis of non-ribosomal peptide synthetases (NRPSs) in *Trichoderma virens* reveals a polyketide synthase (PKS)/NRPS hybrid enzyme involved in the induced systemic resistance response in maize. Microbiology 158:155–165
8. Mukherjee PK, Horwitz BA, Herrera-Estrella A, Schmoll M, Kenerley CM (2013) *Trichoderma* research in the genome era. Annu Rev Phytopathol 51:105–129
9. Brotman Y, Landau U, Cuadros-Inostroza A, Takayuki T, Fernie AR, Chet I, Viterbo A, Willmitzer L (2013) *Trichoderma*-plant root colonization: escaping early plant defense responses and activation of the antioxidant machinery for saline stress tolerance. PLoS Pathog 9:e1003221
10. Saxena A, Raghuwanshi R, Singh HB (2014) *Trichoderma* species mediated differential tolerance against biotic stress of phytopathogens in *Cicer arietinum* L. J Basic Microbiol 55: 195–206

# Part II

## Evaluating Mode of Action of Microorganisms

# Chapter 3

# Methods for the Evaluation of the Bioactivity and Biocontrol Potential of Species of *Trichoderma*

**Johanna Steyaert, Emily Hicks, Janaki Kandula, Diwakar Kandula, Hossein Alizadeh, Mark Braithwaite, Jessica Yardley, Artemio Mendoza-Mendoza, and Alison Stewart**

## Abstract

Members of the genus *Trichoderma* comprise the majority of commercial fungal biocontrol agents of plant diseases. As such, there is a wealth of information available on the analysis of their biocontrol potential and the mechanisms behind their superior abilities. This chapter aims to summarize the most common methods utilized within a *Trichoderma* biocontrol program for assessing the biological properties of individual strains.

**Key words** *Trichoderma*, Biocontrol, Endophyte, Phytopathogens, Induced systemic resistance, Mycoparasitism, Antibiosis, Volatiles, Nematodes

## 1 Introduction

*Trichoderma* species are versatile, ubiquitous filamentous fungi which are found free living in soil, colonizing dead organic matter, and in beneficial endophytic associations with plant species. Collectively, *Trichoderma* species comprise the majority of commercially available fungal agents for the control of plant diseases (caused by fungi, bacteria, and nematodes). Effective biocontrol is achieved through a combination of mechanisms including mycoparasitism, competition for nutrients and/or space, antibiosis, and induction of systemic resistance [1–4]. Additionally, some *Trichoderma* strains induce temperature and drought tolerance [2, 4]; while other strains induce plant growth promotion and effectively act as biofertilizers [5, 6]. In this chapter, we describe diverse methods for assessing the biological properties of *Trichoderma* and their potential use as biopesticides.

---

The original version of this chapter was revised. The erratum to this chapter is available at: DOI 10.1007/978-1-4939-6367-6_17

Travis R. Glare and Maria E. Moran-Diez (eds.), *Microbial-Based Biopesticides: Methods and Protocols*, Methods in Molecular Biology, vol. 1477, DOI 10.1007/978-1-4939-6367-6_3, © Springer Science+Business Media New York 2016

## 2 Materials

### 2.1 General Media

All media are autoclaved at 121 °C for 15 min, cooled to 50 °C, and poured into 90 mm Petri dishes, unless otherwise stated. Standard media [e.g., potato-dextrose agar (PDA) and broth (PDB), malt extract agar (MEA) and broth (MEB) is prepared according to manufacturer's instructions]

1. *Trichoderma* selective medium (TSM): 10.0 g malt extract, 1.0 g yeast extract, 0.2 g Terraclor (quintozene) and 0.15 g Rose Bengal, add 900 mL deionized water (DW) and stir until dissolved. Add 15.0 g agar, and fill up to 1000 mL with DW. Bring to the boil in the microwave (6–7 min) and add 0.6 mL chloramphenicol stock solution (100 mg/mL) before autoclaving. Alternatively add 10 mL of filter-sterilized streptomycin/chlortetracycline solution after autoclaving. Plates should be stored at 4 °C.
2. Antibiotics stock solution for TSM: Weigh 2.5 g streptomycin sulfate and 0.5 g chlortetracycline HCl, add 100 mL of DW and stir until fully dissolved. Filter the solution to 0.2 μm and store aliquots at −20 °C.
3. Malt–Yeast extract (MYE) broth: Suspend 20 g of malt extract and 2 g of yeast extract in 1 L distilled water. Mix thoroughly to dissolve and sterilize by autoclaving.

### 2.2 Nonvolatile Metabolites Production

1. Autoclaved sterile cellophane disks (8 cm diameter).
2. Bio-Rad, Model 583 gel dryer Backing, Cat. 1650963.
3. Plastic cling wrap.

### 2.3 Rhizospheric and Endophytic Colonization of Maize. Modification: Growth Promotion and Systemic Resistance

1. Incubator (23 °C, 12/12 h light/dark).
2. SDW with 0.1 % Tween 80.
3. Miracloth (Merck Millipore).
4. 2 % methylcellulose (BDH GPRTM methyl cellulose).
5. Soil substrate, e.g., "John Innes" mix (field soil, blood and bone fertilizer, superphosphate, potassium phosphate, agricultural lime, dolomite lime).
6. Universal bottles containing 9 mL sterile distilled water (SDW) with 0.1 % Tween 80.
7. TSM plates.
8. Sterile paper towels.
9. 5 % sodium hypochlorite (*see* **Note 1**).

### 2.4 Detached Strawberry Necrotic Leaf Assay

1. Incubator (25 °C, 12/12 h light/dark).
2. SDW with 0.1 % Tween 80.
3. Miracloth (Merck Millipore).

4. Funnels and sterile universal bottles.
5. Half strength PDA plates poured in deep Petri dishes (90 × 25 mm) for spore production of *Botrytis cinerea*.
6. Half strength PDB (½ PDB).
7. Fresh strawberry leaves.
8. 2 % sodium hypochlorite (*see* **Note 1**).
9. DAS® herbicide (simonize 4.8 g L-1, amitrole 1.5 g L-1, 2, 2-dichloropropionic acid 3.1 g L-1(Yates, Homebush, Australia).
10. Mini-humidity chambers (takeaway boxes-170 × 440 × 80 mm).

#### 2.5 Biocontrol Activity Against B. cinerea in Grapes

1. Incubator (25 °C, 12/12 h light/dark).
2. SDW with 0.1 % Tween 80.
3. Miracloth (Merck Millipore).
4. Half strength PDA plates poured in deep Petri dishes (90 × 25 mm) for spore production of *Botrytis cinerea*.
5. Half strength PDB (½ PDB).
6. Rooted cuttings of Pinot noir.
7. *Trichoderma* spp spore suspension ($10^7$ spores/mL).
8. *Botrytis cinerea* spore suspension ($10^5$ spores/mL).
9. Handheld sprayer.

#### 2.6 Wheat/Root Lesion Nematode Bioassay

1. Millipore paper tape.
2. SDW with 0.1 % Tween 80.
3. Miracloth (Merck Millipore).
4. Large plastic container to fit plastic rack in, modified to include a bung so that liquid can be allowed to drain.
5. PVC piping (35 mm diameter) cut into 20 cm lengths.
6. Sand.
7. O-rings.
8. Root lesion nematodes (RLN) (*Pratylenchus* spp.).
9. Overhead mist extraction unit.
10. Doncaster counting dish.

## 3 Methods

#### 3.1 Dual Culture Plate Assay (Mycoparasitism)

1. Predetermine the relative growth rates of the *Trichoderma* strains and the pathogens to be tested. If the growth rates are similar, then test plates can be inoculated simultaneously; if they differ markedly then the plates should be preinoculated with the slower growing fungus 1–3 days accordingly.

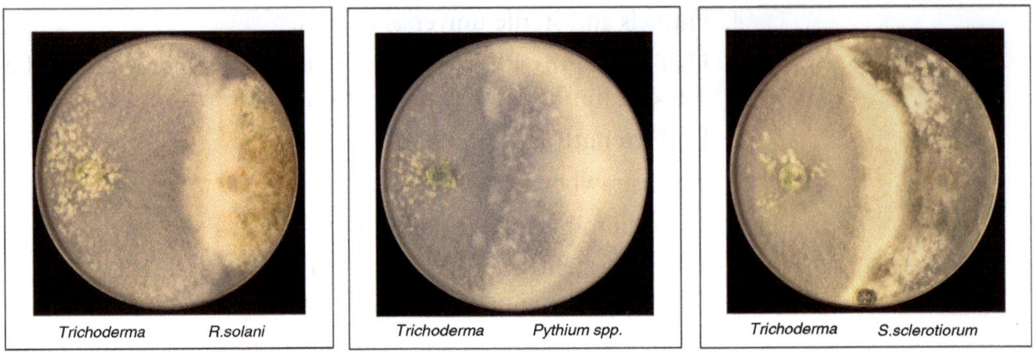


**Fig. 1** Dual culture of *Trichoderma* spp. with different plant pathogens

2. Inoculate PDA plates with a mycelial plug from the colony margin of an actively growing *Trichoderma* culture and to the other side a mycelial plug of the fungal plant pathogen to be tested (e.g., *Rhizoctonia solani, Pythium ultimum, Sclerotinia sclerotiorum, Sclerotium cepivorum*) (*see* **Note 2**) (*see* ref. 7). In addition, inoculate 6× PDA plates each with the pathogen alone (controls)
3. Seal the plates with Millipore tape and incubate at 20 °C under 12/12 h light/dark conditions. Behavior of *Trichoderma* isolates against each pathogen is examined visually until the *Trichoderma* strains have overgrown or surrounded the pathogen colony.
4. Mark the position of the pathogen colony margin on the reverse of the control and test plates daily until growth ceases. At the end of the experiment calculate the growth rates and analyze.
5. Observe daily the morphology of each colony. Record any changes in colony color, pigment secretion, production of conidia or sclerotia, or cessation of colony growth. Photograph the plates at the end of the experiment (Fig. 1).

It is useful to categorize the interaction using predetermined criteria. For example, the interaction between *S. cepivorum* and *Trichoderma* is assessed using the following criteria (*see* refs. 8, 9): **A**, The hyphae of the two colonies intermingle but remain clearly distinguishable. **B**, The growing margins of the two fungi meet; the phytopathogenic fungus is inhibited and overgrown by *Trichoderma*. **C**, The hyphae of the two organisms approach one another and stop growing. **D**, The growth of the phytopathogenic fungus is inhibited at a distance leaving a clear zone of inhibition between the two organisms. Interaction types **B** and **D** were considered to be antagonistic (*see* ref. 9).

### 3.2 Production of Volatile Compounds

The effect of volatile compounds produced by *Trichoderma* spp. on radial growth of pathogenic fungi is determined following the method described by Muthukumar et al. (*see* ref. 7) with some modifications from our group.

1. Inoculate PDA plates (90 × 25 mm) with a mycelial plug from the colony margin of an actively growing *Trichoderma* culture (typically 3 days old) and incubate for 48 h at 23 °C.
2. Inoculate additional PDA plates with a mycelial plug from the colony margin of an actively growing pathogen colony (typically 3–5 days old) (*see* **Note 2**). Replace the lid from the *Trichoderma* culture by the PDA plate inoculated with the plant pathogen so that both fungi are facing each other. The plant pathogenic fungi need to be inverted over *Trichoderma* to avoid contamination by the mycoparasitic fungal conidia. Seal the Petri dishes together with Parafilm M (*see* **Note 3**). Incubate at 23 °C under 12/12 h light/dark conditions for 4 days (Fig. 2).
3. For the control, use an uninoculated PDA plate in place of the *Trichoderma* culture. Each treatment requires at least four replications.
4. Measure the colony diameter of the plant pathogen daily on both controls and treatment plates and calculate radial growth rate.
5. The percentage of inhibition of the plant pathogen over control is calculated using the formula: Mycelial Inhibition $\% = [(C-T)/C] \times 100$. Where, C = radial mycelial growth of the plant pathogen on the control plates (cm) and T = radial mycelial growth of the plant pathogen in the presence of *Trichoderma* (cm) (*see* ref. 10).

### 3.3 Production of Nonvolatile Metabolites

1. Overlay PDA plates with cellophane and centrally inoculate with a mycelial plug from the colony margin of an actively growing *Trichoderma* colony. Seal plates with plastic cling wrap and incubate at 23 °C under 12/12 h light/dark conditions for 3 days.
2. Remove the cellophane and inoculate the same plate with a mycelial plug from the colony margin of an actively growing pathogen colony (typically 3–5 days old) (*see* **Note 2**). The control plates are treated as earlier except the cellophane is left uninoculated.
3. Incubate as earlier until the colony of plant pathogen covers the whole Petri dish. The control treatment contains the plant pathogen grown on an uninoculated *Trichoderma* PDA plate.
4. Measure the colony diameter of the plant pathogen daily on both controls and treatment plates and calculate radial growth rate and calculate mycelial inhibition as described in **step 5**, Subheading 3.2.

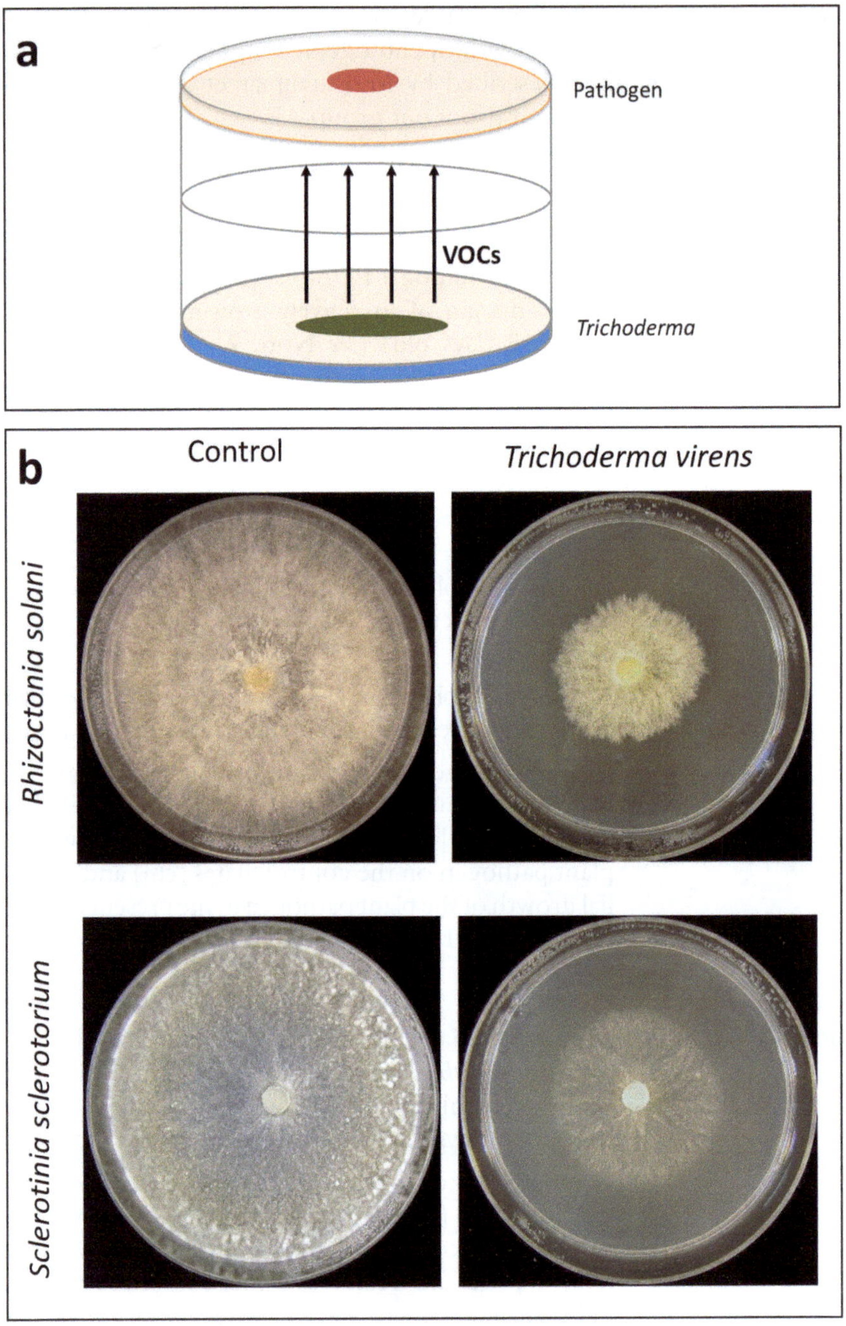


**Fig. 2** Volatile organic compounds emitted by *Trichoderma* and their effect in two plant pathogens. (**a**) Graphic representation of the bio-assay. (**b**) Effect of VOC in *R. solani* (*upper panel*) and *S. sclerotorium* (*lower panel*)

### *3.4 Rhizospheric and Endophytic Colonization of Maize. Modification: Growth Promotion and Systemic Resistance*

1. Prepare *Trichoderma* conidia on PDA plates as described in Chapter 2; **steps 1–3** in Subheading 3.3. Dilute to $5 \times 10^8$ conidia/mL in 0.1% Tween 80 then mix an equal volume with 2% methylcellulose to give a final concentration of $2.5 \times 10^8$ conidia/mL.
2. Treat maize seeds in batches of 5 g with 100 μL of the conidia/methylcellulose solution prepared earlier. Mix till they are dry and evenly covered. To confirm the number of conidia loaded per seed, shake two aliquots of five of the treated seeds in 5 mL of sterile water each on a wrist action shaker for 10 min, and count in a hemocytometer the number of conidia present in the washings. Control seeds are coated with 1% methylcellulose only (CMC control) or not coated at all (bare seed control).
3. For determination of the conidial germination rate, an aliquot of the initial conidial suspension is diluted to $2.5 \times 10^5$ in ½ strength PDB in $2 \times 1.7$ mL centrifuge tubes and incubated on the rotating wheel of a hybridization oven (6 rpm) at 23 °C for 18 h. Place the tubes on ice and record germination for 50 conidia from four samples per tube.
4. The activity of *Trichoderma* is multifactorial, including variations in the soil composition. It is recommended to use standard soil mixtures to control external factors, e.g., "John-Innes" mix (*see* ref. 11). But the growing medium is at the discretion of the researcher and local soil may be used. However, we advise that the physicochemical characteristics of the soil are analyzed. The "John Innes" soil mix is as follows: Sieved field soil is mixed with peat and pumice at a ratio of 7:3:2- v/v/v and amended with 1.2 g/L blood and bone, 1.2 g/L superphosphate, 0.3 g/L potassium sulfate, 2.0 g/L agricultural lime (calcium carbonate), and 3.5 g/L dolomite lime (calcium carbonate). The growing media is mixed thoroughly before use. Determine the soil mix moisture content and adjust to desired level (typically, we adjust to 28%).
5. Fill pots or bags with soil mix and plant one maize seed per bag at a depth of 2.5 cm and place in a glasshouse. Daily temperature highs should be ideally between 20 and 25 °C with 16 h light and 8 h dark. Monitor the soil and water all bags/pots as required every third day.
6. Record emergence percentage after 1 week. Sample times are at the discretion of the researcher; typically we destructively harvest at 1 week, 3 weeks, and 6 weeks after planting.
7. At each sample time, gently remove the plant and remove 1 g of the loosely adhering soil from the top, middle, and bottom

zones of the root and place root aside for endophytic determination (*see* **step 8** in Subheading 3.4). Process each soil sample as described earlier in Subheading 3.2, **steps 3–4** and count the number of *Trichoderma* colonies present on plates which contain between 20 and 200 colonies. A subsample of soil from each zone is combined and moisture content determined. Calculate the CFU (Colony Forming Units) per g of dry soil using the CFU counts. Include the negative controls in the analysis to determine the level of background *Trichoderma* within the experimental system. The negative control should be significantly lower than the treatments in order for the experiment to be valid. If there is no difference then modifications to the preparation and potentially location of the experiment should be explored.

8. Process root samples for endophytic determination using a modification of the method from Kleinfeld and Chet [12]. Excise the roots below the crown, rinse under running tap water to remove soil particles and dry briefly on sterile paper towels. Soak roots in a 5 % sodium hypochlorite solution for 5 min, then rinse three times in SDW for 2 min and place on fresh sterile paper towels to dry. Cut the roots into five equal pieces and place on the surface of TSM. Incubate the plates unsealed at 20 °C, in the dark for 7 days and record the presence of *Trichoderma* mycelium growing out from the roots. To check the quality of surface sterilization plate out the SDW used in the final rinse from every 20th sample. Percent colonization is based on the number of infected root pieces whereby one piece = 20 % and five pieces = 100 %.
9. To assess growth promotion potential, measure the length of the roots and the shoots and determine their dry weights. Compare results to the non-*Trichoderma* controls.
10. To assess the ability of *Trichoderma* to induce systemic resistance, challenge the shoots with a pathogen. Assess disease incidence and compare to non-*Trichoderma* controls (*see* **Note 4**).

### *3.5 Detached Strawberry Necrotic Leaf Assay*

1. Prepare *Trichoderma* conidia as described in Chapter 2; **steps 1–3** in Subheading 3.3. Dilute to $10^7$conidia/mL.
2. Prepare *Botrytis cinerea* conidia (*see* **Note 5**) and dilute to $10^5$ conidia/mL.
3. Surface sterilize the strawberry leaves by soaking in 2.0 % sodium hypochlorite and rinse three times with SDW, and a final wash for 30 s in DAS® herbicide.
4. Treatments are as follows: (A) *Trichoderma/Botrytis*; (B) *Botrytis cinerea* only; (C) 0.1 % Tween 80 only. Spray the sterile leaves with the *Trichoderma* ($10^7$conidia/mL) suspension (A) or with 0.1 % Tween 80 (B) and (C) until near runoff. Maintain

the leaves in a laminar flow until dry. Then challenge the leaves with *Botrytis cinerea* at $10^5$ conidia/mL (A) and (B).

5. After brief drying, place the leaves in mini-humidity chambers (170×440×80 mm) and incubate at 20 °C, 12/12 h light/dark for 7 days.
6. Use a disease rating scale of 1–4 which corresponds to conidiophore coverage of 0–20 %, 21–40 %, 61–80 %, 81–100 %, respectively. Analyze in comparison to the pathogen control (B). The water control (C) is a control for the surface sterilization.

### 3.6 Biocontrol Activity Against *B. cinerea* in Grapes

1. Prepare rooted cuttings of Pinot noir as follows: Collect uniform hardwood cuttings (4–5 nodes in length) in the autumn and plant in plastic crates filled with perlite which are placed on a thermostatically controlled heating pad providing 26 °C to the base of the cuttings. Keep cuttings moist by spraying with water on alternate days. After 30 days plant the cuttings into potting mix amended with slow release fertilizer and transfer to a semishade house, where the experiment will be conducted. Following bud burst remove the proximal and adjacent leaves to encourage fast growth of the inflorescence. Excise the shoot tips as soon as they appear leaving only 4–6 leaves per plant to ensure all nutrients from the stem are diverted to fruit development. Maintain this number of leaves throughout the experiment by regular trimming.
2. Prepare *Trichoderma* conidia as described in Chapter 2; **steps 1–3** in Subheading 3.3 and dilute to $10^7$ conidia/mL.
3. Prepare *Botrytis cinerea* conidia (*see* **Note 5**) and dilute to $10^5$ conidia/mL.
4. Treatments are as follows: (A) *Trichoderma/Botrytis*; (B) *Botrytis cinerea* only; (C) 0.1 % Tween 80 only. For treatment (A), apply *Trichoderma* spp. using a handheld sprayer till near runoff. This treatment is sprayed three times, one at flowering (20 % cap fall), other at 80 % cap fall, third at prebunch closure (Fig. 3a–b).

**Fig. 3** Biocontrol activity of *Trichoderma* in grapes. (**a**) Pinot Noir Rooted cuttings potted. (**b**) Flower bunches. (**c**) Spraying of flower bunches at 20 % cap fall with shot gun

5. When plants are 80% cap fall apply *B. cinerea* spores to treatments (A) and (B). This application has to be only once. For plants treated with *Trichoderma* (A), *B. cinerea* spores are sprayed 24 h after the second *Trichoderma* spray (80% cap fall) (Fig. 3c).
6. For treatment (C), spray with 0.1% Tween 80 at 80% cap fall.
7. Percentage of *Botrytis* colonization on bunches is assessed at maturity. A disease rating scale of 1–4 which corresponding to conidiophores coverage/rotting of bunches as 0–20%, 21–40%, 61–80%, 81–100%, respectively. Compare the weight of bunches. Analyze in comparison to the pathogen control (B). The 0.1% Tween 80 control is to assess whether background diseases have entered the experimental system.

### 3.7 Wheat/Root Lesion Nematode Bioassay to Screen Beneficial Trichoderma Isolates

The Wheat/root lesion nematode bioassay was initially developed by the South Australia Research and Development Institute (SARDI) and modified within our group (*see* ref. 13).

1. Prepare *Trichoderma* conidia on PDA plates as described in Chapter 2; **steps 1–3** in Subheading 3.3. Dilute to $1 \times 10^9$ conidia/mL in SDW. Add 25 μL of each conidial suspension (~$10^6$ conidia per seed) to 1 g of wheat seeds in a standard Petri dish and shake for approximately 30 s until all of the seeds are evenly coated with conidia. Control seeds are coated with SDW and treated as earlier.
2. For determination of conidial germination, *see* Subheading 3.4, **step 3**. Conidia are considered germinated if the germ tube length is at least twice the diameter of the conidia.
3. Fill the PVC piping with 250 g of sand and then cover the base of the tube with a single layer of Miracloth fastened with an O ring and place in a plastic rack within the large plastic container. Plant two wheat seeds per tube at a depth of approximately 0.5 cm and place in a glasshouse at 18 °C with 16 h light and 8 h dark. Water the tubes as required by flooding the large plastic container and drain the excess water by removing the bung after 5 min. Figure 4 shows a typical bioassay layout.
4. The nematodes are extracted from a culture maintained on carrot callous using an overhead misting unit delivering 10 s of mist every 10 min for a period of 96 h. The nematodes are collected, disinfected by washing four times with 1% streptomycin and penicillin, counted, and adjusted to the required inoculation rates using a Doncaster counting dish.
5. Once the seedlings have emerged (~5 days) remove the second wheat seed. Inoculate RLN at a rate of ~1500 nematodes/tube (a rate that significantly reduces plant biomass after 21-day growth).

**Fig. 4** Bioassay layout showing the PVC piping tubes within the racks with wheat seedlings

6. After 21 days, gently removed the plant to be harvested. Lightly shake the plant and the root system to remove loosely adhering soil and wash under tap water. Place in an overhead mist extraction unit for 96 h to extract the RLN from the roots. Measure the length of the roots and the shoots and determine their dry weights.
7. Count the extracted RLN using a Doncaster counting dish.

## 4 Notes

1. Use a common household bleach and check the percent active ingredient (sodium hypochlorite)—typically 3–5 %. Dilute with water to achieve the required sodium hypochlorite percentage.
2. Incubation time depends on the pathogen's growth rate. Ideally the diameter of the fungal colony should be around 6 cm.
3. Parafilm M is preferred to plastic film as it provides an airtight seal therefore reducing inconsistency between replicates.
4. Some *Trichoderma* endophytes will traverse up the plant and into the shoots. To ensure the effect is systemic and not due to the presence of *Trichoderma*, surface sterilize the shoots of nonpathogen treatments as described in **step 8**, Subheading 3.4 for roots and plate onto TSM. Alternately an isolate-specific marker can be employed to check for *Trichoderma* DNA.

5. *Botrytis* spore production: Inoculate half strength PDA in deep plates (25 mm deep) with fresh 3-day-old mycelium at 5 spots at equidistance for maximizing spore production. Incubate for 15 days at 20 °C in the dark. Harvest the spores as described for *Trichoderma* in Chapter 2; **steps 1–3** in Subheading 3.3.

## Acknowledgements

The work of the *Trichoderma* team from Lincoln University has been supported by the Bio-Protection Research Centre, a Centre of Research Excellence and by the following New Zealand funding agencies/sources: Tertiary Education Commission; Foundation for Research Science and Technology; Ministry for Business, Innovation and Employment; Agricultural and Marketing Research and Development Trust; Marsden Research Fund; Lincoln University Research Fund; Lincoln-Massey Research Excellence program. V.M is supported by the Brazilian government to undertake a doctoral stay at Lincoln University. The nematode work was funded by the Australian Government Grains Research and Development Corporation.

### References

1. Calistru C, McLean M, Berjak P (1997) In vitro studies on the potential for biological control of *Aspergillus flavus* and *Fusarium moniliforme* by *Trichoderma* species. A study of the production of extracellular metabolites by *Trichoderma* species. Mycopathologia 137: 115–124
2. Druzhinina IS, Seidl-Seiboth V, Herrera-Estrella A, Horwitz BA, Kenerley CM, Monte E, Mukherjee PK, Zeilinger S, Grigoriev IV, Kubicek CP (2011) *Trichoderma*: the genomics of opportunistic success. Nat Rev Microbiol 9:749–759
3. Mukherjee PK, Buensanteai N, Moran-Diez ME, Druzhinina IS, Kenerley CM (2012) Functional analysis of non-ribosomal peptide synthetases (NRPSs) in *Trichoderma virens* reveals a polyketide synthase (PKS)/NRPS hybrid enzyme involved in the induced systemic resistance response in maize. Microbiology 158:155–165
4. Mukherjee PK, Horwitz BA, Herrera-Estrella A, Schmoll M, Kenerley CM (2013) *Trichoderma* research in the genome era. Annu Rev Phytopathol 51:105–129
5. Brotman Y, Landau U, Cuadros-Inostroza A, Takayuki T, Fernie AR, Chet I, Viterbo A, Willmitzer L (2013) *Trichoderma*-plant root colonization: escaping early plant defense responses and activation of the antioxidant machinery for saline stress tolerance. PLoS Pathog 9:e1003221
6. Saxena A, Raghuwanshi R, Singh HB (2014) *Trichoderma* species mediated differential tolerance against biotic stress of phytopathogens in *Cicer arietinum* L. J Basic Microbiol 55:195–206
7. Muthukumar A, Eswaran A, Sanjeevkumas K (2011) Exploitation of Trichoderma species on the growth of pythium aphanidermatum in chilli. Braz J Microbiol 42:1598–1607
8. Stewart A, Harrison YA (1989) Mycoparasitism of sclerotia of sclerotium cepivorum. Australas Plant Pathol 18:10–14
9. McLean KL (2001) Biological control of onion white rot using Trichoderma harzianum, in soil, plant & ecological division. Lincoln University, Lincoln, p 214
10. Baratta MT et al (1998) Antimicrobial and antioxidant properties of some commercial essential oils. Flavour Fragrance J 13(4):235–244
11. Bunt AC (1976) Modern potting composts: a manual on the preparation and use of growing

media for pot plants. Modern potting composts. 1976

12. Kleifeld O, Chet I (1992) Trichoderma-harzianum - interaction with plants and effect on growth-response. Plant and Soil 144: 267–272
13. Yardley J, Braithwaite M, Marshall J, Wilson M (2013) Development of a wheat/root lesion nematode bioassay to screen beneficial Trichoderma strains. In: 19th Australasian Plant Pathology Conference Auckland, NZ. pp. 121

# Part III

## Mass Production and Formulations: Bacteria

# Chapter 4

# Purification of the *Yersinia entomophaga* Yen-TC Toxin Complex Using Size Exclusion Chromatography

Sandra A. Jones and Mark R.H. Hurst

## Abstract

The *Yersinia entomophaga* toxin complex (Yen-TC) is the bacterium's main virulence determinant. Because of its high insect activity, methods were developed to allow the routine isolation and purification of Yen-TC from an overnight bacterial culture using size exclusion chromatography. Here we outline an overnight purification procedure using a 100-ml culture volume, where approximately 2 mg of Yen-TC, with an approximate purity of 95–98 %, can be routinely obtained.

**Key words** Size exclusion chromatography, *Yersinia entomophaga*, Yen-TC, Purification

## 1 Introduction

The main virulence determinant of *Yersinia entomophaga* [1] is an insect-active toxin complex (TC)-derivative termed the Yen-TC [2]. TCs were first identified in the genome of *Photorhabdus luminescens* [3] and have since been identified in other bacterial genera, including members of the genus *Yersinia* [2]. Typically, TCs are composed of three proteins, TC-A, TC-B, and TC-C, which combine to form the insect-active complex. The Yen-TC is comprised of seven subunit proteins: two TC-A-like proteins (YenA1 and YenA2), a TC-B-like protein (YenB), two TC-C-like proteins (YenC1 and YenC2), and two chitinases (Chi1 and Chi2), which combine to form the insect-active Yen-TC [2]. The final protein has a predicted mass of approximately 2360 kDa. Recent structural analysis has revealed that both the Yen-TC and the *P. luminescens* TC-A (TcdA1) form a pentameric cage [4], which, in the case of the Yen-TC, is decorated with functional chitinases [5, 6]. *Y. entomophaga* produces large amounts of Yen-TC when cultured at 25 °C, but no Yen-TC was detected in either the cell pellet or the culture supernatant at 37 °C [2]. The large size of the Yen-TC makes it conducive to purification by size exclusion chromatography (SEC).

Travis R. Glare and Maria E. Moran-Diez (eds.), *Microbial-Based Biopesticides: Methods and Protocols*, Methods in Molecular Biology, vol. 1477, DOI 10.1007/978-1-4939-6367-6_4, © Springer Science+Business Media New York 2016

## 2 Materials

### *2.1 Items Required for Culturing*

1. Spectrophotometer and microphotometer cuvettes, 2.0-ml.
2. Orbital mixer incubator (Raytek Corporation).
3. 250-ml Erlenmeyer flasks.
4. Centrifuge with rotor capacity for 1.7-ml microcentrifuge tubes and rotor capacity for 15-ml and 50-ml centrifuge tubes.
5. Luria-Bertani (LB) agar and broth.
6. *Yersinia entomophaga* [American Type Culture Collection (ATCC BAA-1678)].

### *2.2 Yen-TC Purification*

1. 0.22-µm pore-size cellulose acetate syringe filters.
2. Ammonium sulfate $(NH_4)_2SO_4$.
3. Centrifugal filtration concentrators: Amicon Ultra-4 Centrifugal Filter Units: Ultracel regenerated cellulose membrane, 30 K and 50 K nominal molecular weight limit (NMWL).
4. Sephacryl S-400 High Resolution size exclusion chromatography (SEC) medium.
5. Glass Econo-Column 1.5 × 50 cm (D × L) with Econo-Column flow adaptor (Bio-Rad).
6. BioLogic LP System, low pressure chromatography instrument, and fraction collector (Bio-Rad Laboratories, model: 731–8350).
7. Polypropylene tubes used for fraction collector (12 × 75 mm, 5-ml).
8. Bio-Rad Protein Assay kit (Bradford).
9. 96-well microplates, flat bottom polystyrene, clear.
10. Microplate reader and data analysis software.
11. Tris buffered saline (TBS;10× buffer) with 0.2 % sodium azide (*see* **Note 1**).

### *2.3 Sodium Dodecyl Sulfate Polyacrylamide Gel Electrophoresis (SDS-PAGE) Components*

1. 30 % acrylamide/Bis solution (37.5:1) (acrylamide:Bis), Bio-Rad (*see* **Note 2**).
2. Ammonium persulfate, 10 % solution in water (*see* **Note 3**).
3. *N*, *N*, *N′*, *N′* tetramethylethylenediamine (TEMED) (Sigma Chemical Company) (*see* **Note 4**).
4. SDS-PAGE running buffer: 0.025 M Tris–HCl, pH 8.3, 0.192 M glycine, 0.1 % SDS (*see* **Note 5**).
5. Resolving gel buffer: 1.5 M Tris–HCl, pH 8.8. Add approximately 150 ml of water to a 500-ml glass beaker. Transfer 54.4 g of Tris base to the beaker. Mix on a magnetic stirrer and adjust pH with HCl (*see* **Note 6**). Transfer to a 500-ml

measuring cylinder and make up to 300 ml with water. Store at room temperature.

6. Stacking gel buffer: 0.5 M Tris–HCl, pH 6.8. Add 60 ml water to a 100-ml beaker. Weigh 6.06 g of Tris–HCl and prepare 100 ml of solution as in the previous step (*see* **Note 7**). Store at room temperature.
7. SDS sample buffer (5×) is prepared as follows: 1.0 ml 0.5 M Tris–HCl (pH 6.8), 0.8 ml glycerol, 1.6 ml 10% (w/v) SDS, 0.4 ml 2-mercaptoethanol, 0.4 ml 1% (w/v) bromophenol blue. Store at room temperature for no longer than 4 weeks
8. Vertical electrophoresis unit and 10 × 8 cm glass plates (Hoeffer SE250 Mini).

## 3 Methods

All solutions, unless otherwise specified, were prepared using distilled deionized Millipore micro-filtered (MilliQ) water. All procedures were performed at room temperature (approximately 22 °C) unless otherwise specified.

### 3.1 Culture and Yen-TC Production

1. Inoculate a 3-ml aliquot of LB broth from a pure culture of *Y. entomophaga* grown on a LB plate (*see* **Note 8**). Incubate for 6–7 h at 30 °C and 250 rpm until an $OD_{600}$ of approximately 0.25 is reached.
2. Inoculate two 50-ml aliquots of LB broth in 250-ml Erlenmeyer flasks with 20 μl of seeding culture (0.04% inoculum). Incubate at 25 °C for 17 h with shaking at 250 rpm in an orbital incubator. This should produce a cell density of approximately $1 \times 10^9$ colony forming units (CFU) $ml^{-1}$, with $OD_{600} = 0.45$
3. Following incubation, allow flasks to incubate at room temperature without shaking for 30 min (*see* **Note 9**). Transfer the standing cultures to two 50-ml centrifuge tubes and pellet the cells by centrifugation at 8000 × *g* for 10 min at 4 °C.
4. Following centrifugation, remove the supernatant and pass through a 0.2-μm syringe filter into fresh tubes. Discard the cell pellets.

### 3.2 Purification of Yen-TC

1. Ammonium sulfate precipitation (70% saturation). A 90-ml aliquot of supernatant will require 40.6 g of $(NH_4)_2SO_4$. Transfer the culture supernatant to a 400-ml glass beaker and place on ice with gentle mixing on a magnetic stirrer. Add small quantities of $(NH_4)_2SO_4$ (approximately 5 g) every 15–20 min while gently mixing for 2 h. After the final addition of $(NH_4)_2SO_4$, continue mixing on ice for at least 30 min to fully equilibrate.

2. Transfer the precipitated solution to 50-ml centrifuge tubes and centrifuge at 10,000 × *g* for 20 min at 4 °C (*see* **Note 10**). Resuspend the pelleted proteins in 7.0 ml of 25 mM TBS and aliquot between 1.7-ml microcentrifuge tubes. Centrifuge at 16,000 × *g* for 6 min at 4 °C (*see* **Note 11**) to pellet undissolved debris. Transfer the pooled clarified supernatants to a clean 15-ml centrifuge tube. Concentrate to a final volume of 0.5 ml using an Amicon 50 K 4.0-ml ultra filter spin concentrator. Centrifuge at 5000 × *g* for 10–15 min using an Eppendorf 5810R centrifuge, fitted with an F34 rotor. Repeated centrifugation steps may be required to achieve the final reduction in volume (*see* **Note 12**).
3. Insert a pipette tip into the bottom of the filter unit (*see* **Note 13**) to collect the concentrated retentate. Transfer to a microcentrifuge tube and then centrifuge at 20,000 × *g* for 5 min to remove undissolved material.
4. Apply the supernatant to a Sephacryl S-400 HR SEC column (1.5 × 46 cm bed volume) (*see* **Note 14**) and elute using 25 mM TBS, at a flow rate of 0.5 ml/min at room temperature. Collect fractions every 1.5 min. Monitor protein elution by measuring absorbance at 280 nm. $OD_{280}$ values of 0.1–1.0 can be expected (Fig. 1) (*see* **Note 15**).

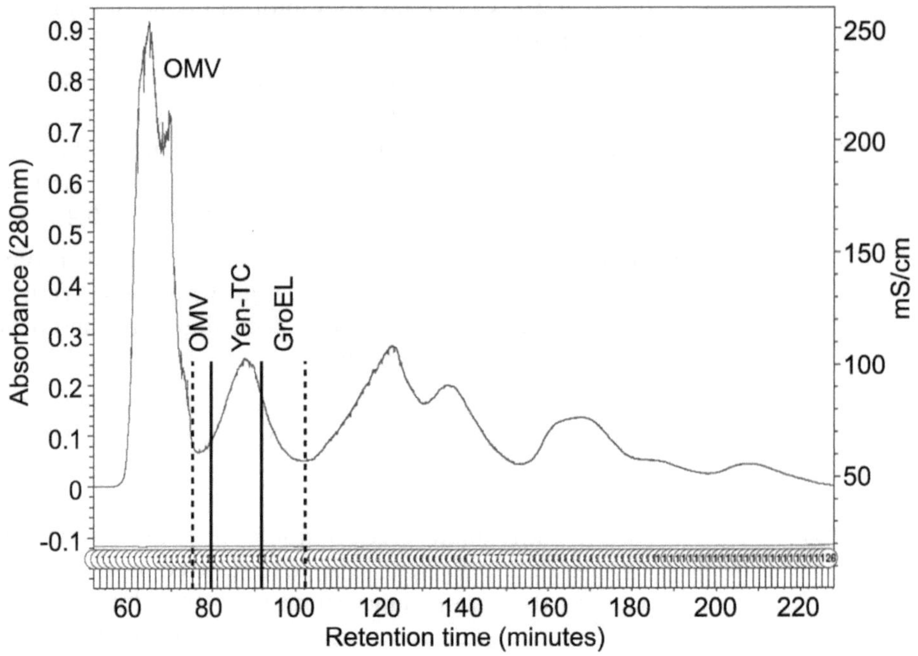


**Fig. 1** Sephacryl S400 SEC profile of purified Yen-TC. The fractions containing Yen-TC at the highest purity are within the *solid vertical lines* (typically fractions 31–35). *Vertical dashed lines* denote the fractions containing Yen-TC with either OMV or GroEL proteins

**Table 1**
**Composition of SDS-PAGE resolving gel**

| % gel (resolving) | 6 | 7.5 | 8 | 10 | 12 |
|---|---|---|---|---|---|
| dd$H_2O$ (ml) | 5.3 | 4.8 | 4.6 | 3.9 | 3.3 |
| 1.5 M Tris–HCl pH 8.8 (ml) | 2.5 | 2.5 | 2.5 | 2.5 | 2.5 |
| Acrylamide/Bis (30%) (ml) | 2 | 2.5 | 2.66 | 3.3 | 4 |
| 10% SDS (μl) | 100 | 100 | 100 | 100 | 100 |
| 10% ammonium persulfate (μl) | 50 | 50 | 50 | 50 | 50 |
| TEMED (μl) | 5 | 5 | 5 | 5 | 5 |

**Table 2**
**Composition of SDS-PAGE stacking gel**

| % gel (stacking) | 4 |
|---|---|
| dd$H_2O$ (ml) | 6.1 |
| 0.5 M Tris–HCl pH 6.8 (ml) | 2.5 |
| Acrylamide/Bis (30%) (ml) | 1.3 |
| 10% SDS (μl) | 100 |
| 10% ammonium persulfate (μl) | 50 |
| TEMED (μl) | 10 |

### 3.3 SDS-PAGE

Fractions containing Yen-TC can be identified by SDS-PAGE analysis on 10% gels using the buffer system of Laemmli [7], and stained with the silver stain described by Blum et al. [8], as outlined as follows.

1. Mix the components of the resolving gel and stacking gel in the order given in Tables 1 and 2 in two separate 15-ml centrifuge tubes and mix gently (*see* **Note 16**). Pour the resolving gel into the gel cassette, allowing space for the stacking gel. Immediately overlay the gel surface with the stacking gel (*see* **Note 17**) and insert a 15-well comb (*see* **Note 18**). Leave to polymerize for 45 min.
2. Pipette into a microcentrifuge tube a 20-μl aliquot from every second or third fraction tube from the expected Yen-TC peak on the chromatograph (Fig. 1). Add 8 μl of 5× SDS sample buffer and heat at 95 °C for 5 min (*see* **Note 19**).
3. Carefully load 20 μl into the gel wells (*see* **Note 20**). Electrophorese at a constant 200 V (~30–40 mA/gel) for

**Table 3**
**Stepwise silver stain procedure**

| Step | Reagent | Volume | Time | *see* Note |
|---|---|---|---|---|
| Fixative | Methanol<br>Acetic acid<br>Formalin<br>$dH_2O$ | 50 ml<br>12 ml<br>50 µl<br>38 ml | >30 min | |
| Wash | 50 % methanol | 100 ml | >5 min | |
| Wash | 50 % methanol | 100 ml | >5 min | |
| Wash | 50 % methanol | 100 ml | >5 min | |
| Pretreat | 0.1 M sodium thiosulfate<br>$dH_2O$ | 0.8 ml<br>100 ml | 1 min | **22** |
| Rinse | $dH_2O$ | 100 ml | 20 s | |
| Rinse | $dH_2O$ | 100 ml | 20 s | |
| Rinse | $dH_2O$ | 100 ml | 20 s | |
| Impregnate | Silver nitrate<br>Formalin<br>$dH_2O$ | 0.2 g<br>75 µl<br>100 ml | 10–15 min | **23** |
| Rinse | $dH_2O$ | 100 ml | 20 s | |
| Rinse | $dH_2O$ | 100 ml | 20 s | |
| Develop | Sodium carbonate<br>Formalin<br>0.1 M sodium thiosulfate<br>$dH_2O$ | 12 g<br>50 µl<br>16 µl<br>100 ml | 0–10 min | 24<br>**25** |
| Wash | $dH_2O$ | 100 ml | 10 s | |
| Stop | 10 % acetic acid<br>$dH_2O$ | 10 ml<br>100 ml | 10 min | **26** |

60 min or until the bromophenol blue dye front has reached the bottom of the gel.

4. Turn off the power supply and remove the gel from the glass plates into a staining tray (*see* **Note 21**). Proceed with the silver staining procedure outlined in Table 3.
5. Pool fractions containing Yen-TC (Fig. 2) with the lowest concentration of outer membrane vesicles (OMVs) and GroEL proteins (typically Yen-TC fractions elute at 80–105 min). Concentrate 10-fold using an Amicon Ultracel-30 K device. Determine the concentration of purified Yen-TC using a Bio-Rad

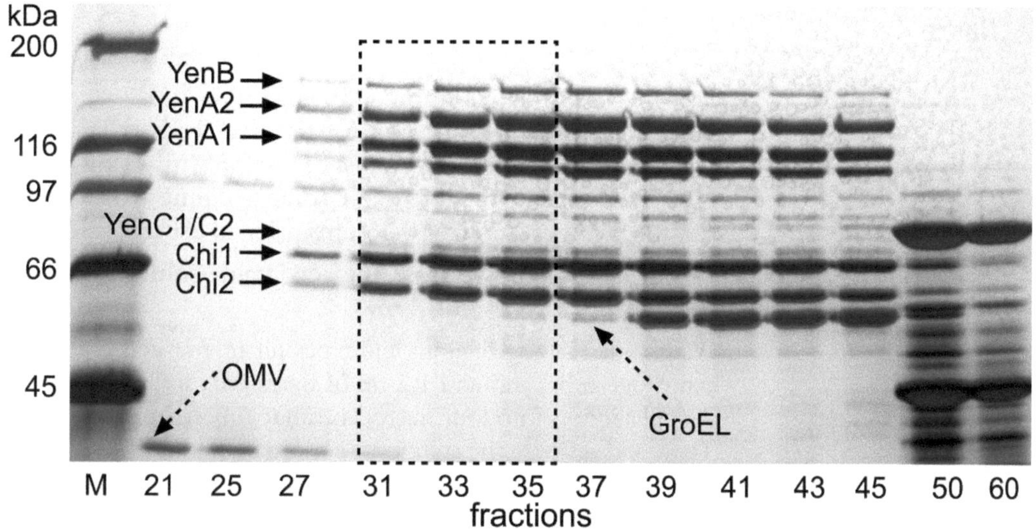


**Fig. 2** Silver stained SDS-polyacrylamide gel of column fractions. The lanes within the *dashed rectangle* are the fractions containing toxin and the lowest concentration of outer membrane vesicles (OMV) (37 kDa) and GroEL proteins (60 kDa). These bands correspond to the Yen-TC peak in Fig. 1 (retention time 80–90 min). These fractions should be pooled and concentrated. The individual toxin components are indicated with *solid arrows* and have been labeled accordingly

Protein Assay (based on the method of Bradford) using bovine serum albumin (BSA) as the standard.

### 3.4 Protein Silver Stain for SDS-PAGE Gels

Procedure is described in Table 3.

Gloves must be worn throughout the staining procedure. All solutions, including the washes, use distilled deionized microfiltered (MilliQ) water. All solutions should be prepared fresh; however, the fixative, 50 % methanol wash solutions, and 0.1 M sodium thiosulfate stock solution can be prepared in advance and stored for later use.

### 3.5 Protein Quantification Using a Bradford Microtiter Plate Assay

The Bradford protein assay [9] is used to measure the protein concentration of purified Yen-TC solution. Comparison to a standard curve made with BSA provides a relative measurement of the protein concentration (*see* **Note 21**).

1. Follow the instructions for microtiter plate assay in the manual provided with the assay kit.
2. Measure the absorbance at 595 nm in a Fluostar Omega microplate reader.
3. Analyze the sample measurements using Omega MARS software.

The expected yield of pure Yen-TC from 100 ml of LB broth culture supernatant is approximately 1.0–2.0 mg.

## 4 Notes

1. Prepare 10× stock solution (250 mM TBS pH 7.4): weigh 40 g NaCl, 1.0 g KCl, 2.42 g Tris base, 16.5 g Tris–HCl. Dissolve in 450 ml $dH_2O$. Adjust pH to 7.5 with ~1 ml of 6 M HCl. Add $dH_2O$ to 500 ml. Use at a dilution of 1 in 10 (25 mM). Filter sterilize (0.45 μm) prior to using for SEC.
2. In our laboratory, we store the acrylamide solution at room temperature.
3. Weigh out 120 mg of ammonium persulfate, transfer to a microcentrifuge tube, and add 1.2 ml of water. Mix well and dispense 300-μl aliquots into four microcentrifuge tubes. Store at −20 °C. Thaw and refreeze the solution no more than four times.
4. TEMED has a strong pungent smell. Store at 4 °C and USE IN A FUMEHOOD.
5. Prepare a 10× stock solution of SDS-PAGE running buffer. DO NOT alter the final pH of the prepared solution. If all ingredients are correctly measured, the pH should be within the correct range of pH 8.3±0.5. To prepare a working solution, dilute 50 ml of concentrated stock solution in 450 ml of water.
6. Typically ~20 ml 6 N HCl is required to alter the starting pH to the required pH 8.8.
7. Typically ~10 ml 6 N HCl is required to alter the starting pH to the required pH 6.8.
8. The most efficient way to do this is using a sterile 200-μl pipette tip attached to a pipette. Gently take a small scrape of colonies with the pipette tip and eject into the 3-ml of LB broth.
9. Leaving the culture flask stationary for at least 30 min results in 1.5× more toxin in the culture supernatant than processing the culture immediately.
10. $(NH_4)_2SO_4$ solution is approximately 5% heavier than water. Therefore, tubes should be weighed prior to centrifugation if a tube containing water is required as a balance.
11. Centrifuging the supernatant will remove any aggregated outer membrane vesicles (OMV) proteins, which will reduce blocking of the spin concentrators in the next step of Yen-TC concentration.
12. At each centrifugation, add ~0.5 ml of TBS to the retentate and mix thoroughly to ensure good buffer exchange and removal of $(NH_4)_2SO_4$. Avoid touching the membrane with the pipette tip.
13. The final 0.5 ml of concentrated retentate can be recovered by inserting a pipette tip into the bottom of the filter unit and withdrawing the sample. Take extra care to wash any deposited

protein from the sides of the tube by gently running TBS (2 × 100 μl) down the membrane several times, and eventually adding to the concentrated retentate. Avoid generating bubbles and foam as this can denature the proteins.

14. The column load volume should be kept to a minimum (<1.0 ml), as larger volumes broaden the eluted peaks.
15. To prepare the column for reuse, run 30 ml of TBS through the column following completion of the run.
16. Avoid introducing air bubbles into the gel solution as this interferes with polymerization.
17. Gently run the stacking gel solution down the inside of the glass plate using a disposable Pasteur pipette.
18. Insert the comb into the stacking gel on an angle to avoid trapping bubbles.
19. Centrifuge the heated samples at 1000 × *g* for 30 s to bring down the condensate.
20. Rinse the gel wells several times with $dH_2O$ to remove any unpolymerized acrylamide before loading samples.
21. Separate the glass plates by gently easing out the gel spacer a small distance and twisting slightly. Use the gel spacer to slide under the gel to separate from the glass surface.
22. Critical times are in italics and should be closely followed.
23. Silver nitrate can stain bare skin and bench surfaces. Silver nitrate solution <u>should not</u> be discarded directly down the drain. Transfer the solution into a large waste vessel with approximately 2 g of NaCl per 100 ml of $AgNO_3$ solution and leave overnight for insoluble $AgCl_2$ to precipitate. The waste solution can then be filtered through Whatman No. 1 filter paper and the liquid can be safely disposed of down the drain. The filter paper can be disposed of following the laboratory's chemical waste procedure.
24. The developing solution requires rapid stirring (on a magnetic stirrer) when adding the sodium carbonate solution to the water.
25. The development stage needs to be monitored carefully to avoid overstaining the gel. Prepare to stop development when bands are still slightly understained by discarding the developing solution and immediately adding the water rinse. Band development will continue until the gel is in the acetic acid stopping solution.
26. Gels can be stored in $dH_2O$ at room temperature for 3–4 days.
27. Use a linear range of protein concentration for the microtiter plate assay of 0.05–0.4 mg/ml, and prepare six dilutions of BSA standard within this range. Make all dilutions in TBS. Assay protein solutions in triplicate.

## Acknowledgement

This work was supported by the New Zealand Foundation for Research, Science, and Technology (C10X0804).

### References

1. Hurst MRH, Becher SA, Young SD, Nelson TL, Glare TR (2010) Yersinia entomophaga sp. nov. isolated from the New Zealand grass grub Costelytra zealandica. Int J Syst Evol Microbiol 61:844–849
2. Hurst MRH, Jones SA, Tan B, Harper LA, Glare TR (2011) The main virulence determinant of *Yersinia entomophaga* MH96 is a broad host range insect active, toxin complex. J Bacteriol 193:1966–1980
3. Bowen D, Rocheleau TA, Blackburn M, Andreev O, Golubeva E, Bhartia R, ffrench-Constant RH (1998) Insecticidal toxins from the bacterium Photorhabdus luminescens. Science 280:2129–2132
4. Busby JN, Panjiker S, Landsberg MJ, Hurst MRH, Lott SJ (2013) The BC component of ABC toxins is an RHS-repeat-containing protein encapsulation device. Nature 501:547–550
5. Landsberg MJ, Jones SA, Rothnagel R, Busby JN, Marshall SDG, Simpson RM, Lott JS, Hankamer B, Hurst MRH (2011) 3D structure of the *Yersinia entomophaga* toxin complex and implications for insecticidal activity. PNAS USA 108:20544–20549
6. Busby JN, Landsberg MJ, Simpson R, Jones SA, Hankamer B, Hurst MR, Lott JS (2012) Structural analysis of Chi1 chitinase from Yen-TC: the multisubunit insecticidal ABC toxin complex of *yersinia entomophaga*. J Mol Biol 13:359–371
7. Laemmli UK (1970) Cleavage of structural proteins during the assembly of the head of bacteriophage T4. Nature 227:680–685
8. Blum H, Beier H, Gross HJ (1987) Improved silver staining of plant proteins, RNA and DNA in polyacrylamide gels. Electrophoresis 8:93–99
9. Bradford MM (1976) A rapid and sensitive method for the quantitation of microgram quantities of protein utilizing the principle of protein-dye binding. Anal Biochem 72: 248–254

# Chapter 5

# Coated Solid Substrate Microbe Formulations: *Pseudomonas* spp. and Zeolite

**Craig R. Bunt, Sally Price, John Hampton, and Scott Stelting**


## Abstract

Formulation provides a means to stabilize for storage and delivery biocontrol and bioremediation agents based on microbes such as bacteria and fungi. Typically it is difficult to both stabilize and deliver fragile non-spore-forming bacteria. Fungal spores might intuitively appear to be easy to stabilize; however, their tendency to germinate in low moisture environments presents challenges for the formulation scientist. Here we present a light background regarding issues with formulating microbes and strategies to help overcome instability and delivery issues.


**Key words** Coating, Formulation, Polymer, Bacteria, Spore

## 1 Introduction

The use of microbial agents is driven by the emergence of new or expanded restrictions placed upon current chemical control agents [1]. Microbe-based formulations have applications ranging from biocontrol and bioremediation to increased plant establishment and persistence. However, microbial-based products often display stability and delivery challenges [2]. Such challenges can be overcome, or partially addressed, by formulation [3, 4]. Even if stability is not an issue, for example, as with spore-forming bacteria like *Bacillus* spp. or *Streptomyces* spp. [5–7], the need for effective delivery may present a limitation requiring formulation to address.

This chapter aims to present a very brief introduction to formulation of microbes and provides details for two broad formulation types: liquid inoculum and dry inoculum coating onto solid substrates. Methods of production, products, and types of formulation are perhaps as varied and numerous as there are potential microbes for the wide ranging applications mentioned earlier. For example, in 2007, a review of myco-insecticides and myco-acaricides identified 171 products worldwide [8].

Travis R. Glare and Maria E. Moran-Diez (eds.), *Microbial-Based Biopesticides: Methods and Protocols*, Methods in Molecular Biology, vol. 1477, DOI 10.1007/978-1-4939-6367-6_5, © Springer Science+Business Media New York 2016

A candidate microbe agent may be identified as being suitably specific (or broad acting), virulent, or the sole microbe identified as capable of achieving a desired outcome, but is initially constrained by a lack of robustness. This challenge can often be overcome or at least suitably addressed by formulation. While at first inspection formulation may simply appear to be the act of combining and suitably mixing a list of ingredients, the need to understand how each component of the mixture may interact together and, most importantly, with the formulated microbe, cannot be ignored [9–16]. However, to date analytical methods that can provide understanding of the mechanisms by which microbes interact with various materials leading to either stabilization or microbe death have yet to be suitably developed and validated.

The term stability encompasses a wide range of meanings, including stability during production and collection of the microbe, stability during processing or compounding of a product, stability during storage, and stability during use. The key formulation stability indicator is microbe viability; this requires that the microbe can be cultured in order to enumerate. Failure to recover viable organisms from a formulation may not indicate a loss of viability, however. In the absence of enumeration, an activity indicating bioassay could be employed and this should include at least some form of standardization in terms of activity. In the absence of a method to enumerate a microbe, bioassays are very difficult to validate. Therefore, enumeration is necessarily by far the most common, to the extent of being almost exclusive, means to measure the quality of microbes that have been produced and compounded as a final product formulation.

### 1.1 Factors Reducing Viability

Once formulated a number of factors may lead to loss of viability, such as dehydration, heat inactivation, excessive moisture, ultraviolet (UV) radiation, and the presence or absence of oxygen. The mechanisms by which many of these factors may lead to a loss in viability are well known, but for some of them the mechanisms are still unclear. Membrane damage of *Lactobacillus plantarum* has been shown to be caused by dehydration but not thermal inactivation [17], while oxidation of *L. bulgaricus* cell membrane lipids has been reported to be proportional to the unsaturated/saturated fatty acid ratio [9]. Many physical conditions can be avoided or minimized in order to maintain or at least improve microbe survival. Obviously heat and UV during compounding should be minimized or avoided, while exposure to oxygen also might need to be controlled for obligate anaerobes. Less well understood is the importance of moisture and desiccation. Microbe stability can be influenced by processing (product manufacturing methods) and final product moisture content and is often highly dependent upon species and sometimes even strain [9, 18–20].

Ultimately formulation aims to minimize any difficulties or limitations associated with handling in order to maintain the viability of a microbe during storage, delivery, and use. A simple formulation method may be to freeze, lyophilize, or spray dry the microbe culture, accepting there will most likely be a loss of microbe viability in the order of 1–5 logs depending on the microbe and method. This might be acceptable, and following appropriate thawing of the product or dilution with an appropriate vehicle, such as chlorine-free water, direct spray application might be sufficient to deliver the biocontrol agent [2].

The relationship between temperature and water activity (a dimensionless quantity representing the energy status of the water in a system and defined as the vapor pressure of water above a sample divided by that of pure water at the same temperature) on survival of a freeze-dried *Lactobacillus* spp. stored in vacuum-sealed foil laminate bags has been reported [2]. As temperature or water activity increased, survival was reduced. For example, at 25 °C, as water activity increased from 0.24 to 0.34, survival reduced from 35 to 25%, respectively, over 6 months. Conversely, at 0.24 water activity, as temperature increased from 25 to 37 °C, survival reduced from 35 to 20%, respectively. At 37 °C and 0.34 water activity, survival was only 10% over 6 months. The relationship between temperature and water activity should be identified for microbes displaying stability issues.

### 1.2 Adsorption on Solid Surfaces

These limitations associated with powder preparations (and particularly their suspension in water to facilitate spray application) can be addressed by adsorption onto a solid substrate, for example, a zeolite granule or a plant seed. Solid substrates can be used to not only improve microbe survival/stability but facilitate their application. A solid substrate may be applied by broadcasting over a field or drilling into a field. Once a microbe is applied to a surface it faces a new challenge as desiccation may lead to a rapid decline in viability. This means that formulations also need to provide in-use stability. For example, a *Lactobacillus* spp. culture on a glass slide that is left to desiccate under normal conditions will lead to complete cell death within 24 h. By incorporation in a gel formulation, viability can be improved. The aim of formulation is to provide a balance between the extremes of freezing and desiccation in order to achieve maximum microbe stability. Avoiding these extremes may not always be easily achieved or can require a compromise in other parameters that affect the yield and viability of microbes [2].

Stabilization of *Serratia entomophilia* (Enterobacteriaceae) has been reported [10]. This work ultimately led to the development of Bioshield™, a product for the control of New Zealand grass grub (*Costelyta zealandica*) that stabilizes *S. entomophilia* by coating it onto zeolite granules. It is worthwhile comparing Bioshield to an earlier product it replaced, Invade®, which was developed in the late 1980s and early 1990s. Invade® was a liquid applied using a modified seed drill at a rate of 1 l/ha diluted to 100 l with non-chlorinated water to deliver $4 \times 10^{13}$ bacteria/ha. It required storage at 4 °C as at 20 °C stability was less than 7 days. Bioshield™, on the other hand, is a granule applied at 30 kg/ha using a seed drill, delivering $4 \times 10^{13}$ bacteria/ha and with stability at 20 °C of 180 days. It has recently been identified that the source or type of zeolite onto which a microbe is coated may be of importance and should be carefully screened to eliminate those that are detrimental to microbe survival. Stelting et al. [16] compared *Pseudomonas* sp. strain ADP survival coated onto two types of zeolite from different sources (a New Zealand and an Australian quarry). The bacteria immobilized onto Australian zeolite remained viable within 1 log unit of initial cfu/g loading and retained their ability to degrade atrazine (as measured by zone clearance on atrazine containing plates) for at least 10 weeks at 25 °C. However when coated onto New Zealand zeolite, viability after 3 weeks storage at 25 °C was reduced by greater than 3 log units of the initial loading.

The term delivery encompasses the process from the time the product is prepared for use, its actual use, or application and the minimum time period after application during which it is required to be active. Broadly, formulations can be either defined as one of two types: solid or liquid. Solid formulations rely upon a physical construct to stabilize and provide delivery. The biocontrol microbe will be either homogenously dispersed through a monolithic matrix or located as a discrete layer in or on a substrate. It is now emerging that a microbe's interaction with its surrounding is one mechanism by which stability can be achieved [21, 22].

An often overlooked aspect of formulation is that once the product is formulated, the ingredients used may make characterization or analysis of microbes difficult. *Lactobacillus* spp. dispersed in a simple gel formulation may be difficult to identify under the microscope and over time this may become more problematical, particularly if trying to distinguish between live and dead microbes and background artifacts due to components of the formulation [2]

### 1.3 Other Coating Techniques

There are many papers and reports [3, 4, 14] describing alginate encapsulation or variations of this type of formulation. The technique appears simple, involving a suspension of a microbe in a solution of sodium alginate that is formed into droplets or emulsified. Adding this to a solution containing polyvalent ions (usually $Ca^{2+}$) induces gelling by cross linking of the alginate, thereby forming a semisolid bead. This bead may be hardened further by the addition of polyvalent cationic polymers (e.g., polylysine or chitosan). Additional processing may also be included, e.g., $Ca^{2+}$ can be extracted from the interior of the hardened bead to liquefy the core (usually using citrate), leaving the hardened shell containing free microbes. While this method (or modifications of) can be used to produce beads or hollow shells, and there are numerous reports using alginates for this purpose, it has met with very little commercial success (if any) and attempts to scale-up the process have tended to encounter process and cost problems [4].

There are many methods to produce coatings or monolithic matrices, such as pelletization, granulation, and film coating. Some techniques require a seed or core on to which layers are applied, allowing for a construct to be "put together" from base materials into the final form. Using a seeding core offers a number of advantages and can be used to dictate and control the final shape and mechanical properties. Along with maintaining microbe viability, solid formulations must also be suitable for delivery and key to this is mechanical robustness (or suitable friability) to ensure that during delivery (e.g., a seed drill or spreader) the formulation does maintain its integrity [7].

Often samples are produced close to or immediately prior to field testing. This test product might not be the same after weeks or months of storage. There is clear evidence now that excipients (the nonactive ingredients of a formulation) play a major role influencing the storage stability of microbes. It is not always clear whether this is due to chemical (e.g., sample pH or ionic strength may be microbicidal), physical (e.g., changes in moisture content may lead to microbe death), or processing properties (e.g., the fermentation method may have produced a microbe with poor stability) and the contributing factors may be difficult to identify. At the very least, samples of test product should be retained under normal conditions until at least the end of field testing and then analyzed for microbial content.

## 2 Materials

For the purpose of illustrating suitable materials used to coat an inoculum onto a solid substrate, the materials of Stelting et al [16] and [23] for *Pseudomonas* sp strain ADP onto zeolite have been selected by way of example. This method applies with obvious modifications for small amounts (10–500 g) of material which may be coated in a suitable beaker and tumbled using a spatula to large (500 g–20 kg) batches that may be coated with the use of a blender such as small vessel roller or fixed blade rotating vessel.

1. Miller's Luria-Bertani (LB) base broth.
2. Glycerol.
3. Xanthan gum from Danisco, China.
4. Extra virgin olive oil as might be purchased from a local food supply retailer.

5. Flowable Atrazine™ (500 g/L atrazine and 50 g/L ethylene glycol) Nufarm NZ Limited, product number 50979-5L.
6. *Pseudomonas* sp strain ADP (DSM 11735), German Collection of Microorganisms and Cell Cultures (DSMZ, Germany) freeze-dried culture.
7. 2–6 mm zeolite granules Clinoptilolite with minor amounts of Mordenite (Zeolite Australia, Australia). The cation exchange capacity and hardness of the zeolite has been reported by the supplier [24] as 120 $cmol_c$/kg and 7 Mohs, respectively.
8. 1 M phosphate buffer solution pH 7.4.

## 3 Methods

The following methods have been reported for culturing and coating *Pseudomonas* sp strain ADP onto Zeolite [16, 23, 25]. However, it is likely that this method will be suitable for most liquid cultures. Culture methods are simply given by way of an example. More suitable or specific culture methods may of course be substituted for other microbes.

### 3.1 Strains Maintenance and Growth

1. Revive the culture by plating onto atrazine agar (1000 mg/L) and incubating at 25 °C.
2. Using a single colony, inoculate a 250 mL flask containing 100 mL of 100 ppm atrazine liquid medium.
3. After 72 h (25 °C, 150 rpm) enumerate the viable cell density by dilution and plating on LB agar. Enumerate viable cell counts in duplicate by serial dilution using phosphate buffer solution (0.1 M). Take triplicate samples of 10 μl from the dilution tubes and plate onto LB agar using the tilt plate technique [26] as follows, place the sample approximately 1 cm from the edge of the agar and tilt the plate to an angle at with the droplet will run across the surface of the agar – stopping approximately 1 cm from the corresponding edge of the agar. Incubate plates at 30 °C for 24 h prior to counting.
4. Harvest cells by centrifugation (10 g, 15 min) and resuspend in LB solution containing 40% (v/v) glycerol. Store cells as 100 μL aliquots in 1 mL micro-centrifuge tubes at −80 °C to serve as the source of culture stock for all subsequent methods.
5. Prepare precultures by resuspending a culture stock contained in a micro-centrifuge tube using 1 mL from a vial containing 15 mL of sterile LB broth and returning the entire contents to the vial. Incubate the vial by shaking at 30 °C and 200 rpm. Harvest the preculture after 18 h and in a 500 mL flask containing 100 mL of LB broth inoculated with 1 mL (1%, v/v)

of preculture. Enumerate cells from the flask after 24 h growth on a shaker (200 rpm) at 30 °C as described in Subheading 3.1, **step 3**.

### 3.2 Coating Zeolite (*See* Note 1)

1. A sample of culture (*see* Subheading 3.1) (*see* **Note 2**) with 4% (w/w) each of xanthan gum (*see* **Note 3**) and olive oil (*see* **Note 4**) is applied onto zeolite at a ratio of 4:96 (*see* **Note 5**) to zeolite with the aid of gentle tumbling to distribute the material over the zeolite. Before adding to the culture the xanthan is wetted with the olive oil and then to this mixture the culture is added with gentle continuous stirring until a paste-like consistency is achieved.
2. Once the culture is applied to the zeolite allow the coated material to air dry overnight at room temperature.

### 3.3 Evaluating Efficacy

1. To assess the amount of immobilization and efficiency, or survival over time, add a 1 g sample of the coated zeolite to 9 g phosphate buffer (0.1 M) followed by serial dilution.
2. Efficiency is calculated as the percent (%) cfu/g enumerated compared to the predicted cfu/g based on the known cfu/ml of the culture and rate applied to zeolite.
3. Survival is calculated as the percent (%) cfu/g at time T compared to the cfu/g at the time of sample preparation.

## 4 Notes

1. Zeolite may be substituted with any desired material that requires coating with a culture. For example, the method described in Subheading 3.2 applies equally to the coating of seeds and food cereals.
2. Freeze-dried culture or dried spores may be used to substitute for liquid culture. If this is the case then disperse sufficient freeze-dried material to achieve the desired cfu/g loading over the material being coated with gentle tumbling. Following this apply a hydrocolloid solution (4%, with our without oil) to bind the freeze-dried culture to the material. Adjustments as required can be made for coating freeze-dried material along the same lines as for coating using liquid cultures.
3. Xanthan may be substituted with other food gums such as guar or gellan or other similar hydrocolloid.
4. Olive oil may be substituted with another vegetable oil; it functions as a process aid for xanthan and it may be possible to omit it. If it can be omitted this will be obvious when compounding, otherwise it may be best to include oil in the formulation. Hydrocolloids (and oil) may be omitted and cultures applied directly onto the material for coating.

5. The ratio of culture to material being coated must be small enough so as to not saturate the material, i.e., ensure the culture coats the material and does not separate or disintegrate fragile materials. The ratio of 4:96 culture:material has typically been found to be suitable for most coating needs. The amount of culture added to material to be coated might need to be increased to a point where liquid culture separates the material; should this happen binders may be applied to help adhere the culture to the material. For example, if using a ratio of 20:80 culture:material then add sufficient talc (approximately equal weight to the culture) while under continued tumbling until a dry granular appearance is achieved. This may take some adjustment of ratios and tumbling to achieve the desired result.

## References


1. Glare T, Caradus J, Gelernter W, Jackson T, Keyhani N, Kohl J, Marrone P, Morin L, Stewart A (2012) Have biopesticides come of age? Trends Biotechnol 30:250–258
2. Bunt C, Swaminathan J (2010) "Formulation of plant protection microbes". In: Zydenbos SM, Jackson TA (eds) "Microbial products: exploiting microbial diversity for sustainable plant production" New Zealand Plant Protection Society, ISBN 978-0-473-16531-4, pp. 75–82
3. Lewis JA, Papavizas GC (1991) Biocontrol of plant diseases: the approach for tomorrow. Crop Prot 10:95–105
4. Bashan Y (1998) Inoculants of plant growth-promoting bacteria for use in agriculture. Biotechnol Adv 16:729–770
5. Emmert EAB, Handelsman J (1999) Biocontrol of plant disease: a (Gram-) positive perspective. FEMS Microbiol Lett 171:1–9
6. Navon A (2000) *Bacillus thuringiensis* insecticides in crop protection – reality and prospects. Crop Prot 19:669–676
7. Brar SK, Verma M, Tyagi RD, Valéro JR (2006) Recent advances in downstream processing and formulations of *Bacillus thuringiensis* based biopesticides. Process Biochem 41:323–342
8. de Faria MR, Wraight SP (2007) Mycoinsecticides and mycoacaricides: a comprehensive list with worldwide coverage and international classification of formulation types. Biol Control 43:237–256
9. Castro HP, Teixeira PM, Kirby R (1995) Storage of lyophilized cultures of *Lactobacillus bulgaricus* under different relative humidities and atmospheres. Appl Microbiol Biotechnol 44:172–176
10. Johnson VW, Pearson JF, Jackson TJ (2001) Formulation of *Serratia entomophilia* for biological control of grass grub. N Z Plant Prot 54:125–127
11. Zhang W, Wolf TM, Bailey KL, Mortensen K, Boyetchko SM (2003) Screening of adjuvants for bioherbicide formulations with *Colletotrichum* spp. and *Phoma* spp. Biol Control 26:95–108
12. Rajkumar M, Lee KJ, Freitas H (2008) Effects of chitin and salicylic acid on biological control activity of *Pseudomonas* spp. against damping-off of pepper. S Afr J Bot 74:268–273
13. Wiyono S, Schulz DF, Wolf GA (2008) Improvement of the formulation and antagonistic activity of *Pseudomonas fluorescens* B5 through selective additives in the pelleting process. Biol Control 46:348–357
14. Dubey SC, Bhavani R, Singh B (2009) Development of Pusa 5SD for seed dressing and Pusa Biopellet 10G for soil application formulations of *Trichoderma harzianum* and their evaluation for integrated management of dry root rot of mungbean (*Vigna radiata*). Biol Control 50:231–242
15. Terefe M, Tefera T, Sakhuja PK (2009) Effect of a formulation of *Bacillus firmus* on root-knot nematode *Meloidogyne incognita* infestation and the growth of tomato plants in the greenhouse and nursery. J Invertebr Pathol 100:94–99
16. Stelting S, Burns RG, Sunna A, Visnovsky G, Bunt CR (2012) Immobilization of pseudomonas

sp. strain ADP: a stable inoculant for the bioremediation of atrazine". Appl Clay Sci 64:90–93. doi:10.1016/j.clay.2011.12.006
17. Lievense LC, Verbeek MAM, Noomen A, van't Riet K (1994) Mechanism of dehydration inactivation of *Lactobacillus plantarum*. Appl Microbiol Biotechnol 41:90–94
18. Vaamonde G, Chirife J (1986) Effect of phosphate buffer on *Staphylococcus aureas* growth at a reduced water activity. Int J Food Microbiol 3:51–55
19. Champagne CP, Mondou F, Raymond Y, Roy D (1996) Effect of polymers and storage temperature on the stability of freeze-dried lactic acid bacteria. Food Res Int 29:555–562
20. Kurtmann L, Carlsen CU, Risbo J, Skibsted LH (2009) Storage stability of freeze-dried *Lactobacillus acidophilus* (La-5) in relation to water activity and presence of oxygen and ascorbate. Cryobiology 58:175–180
21. Bjerketorp J, Håkansson S, Belkin S, Jansson JK (2006) Advances in preservation methods: keeping biosensor microorganisms alive and active. Curr Opin Biotechnol 17:43–49
22. Schoug A, Olsson J, Carlfors J, Schnürer J, Håkansson S (2006) Freeze-drying of *Lactobacillus coryniformis* Si3--effects of sucrose concentration, cell density, and freezing rate on cell survival and thermophysical properties. Cryobiology 53:119–127
23. Stelting S, Burns RG, Sunna A, Bunt CR (2014) Survival in sterile soil and atrazine degradation of pseudomonas sp. strain ADP immobilized on zeolite. Biorem J 18:309–316. doi:10.1080/10889868.2014.938723
24. Zeolite, 2010. http://www.zeolite.com.au/products/technicalspecs.pdf (visited 30 Nov 2010)
25. Mandelbaum RT, Allan DL, Wackett LP (1995) Isolation and characterization of a Pseudomonas sp. that mineralizes the s-triazine herbicide atrazine. Appl Environ Microbiol 61:1451–1457
26. Carvalhal MLC, Oliveira MS, Alterthum F (1991) An economical and time saving alternative to the most-probable-number method for the enumeration of microorganisms. J Microbiol Methods 14:165–170

# Part IV

## Mass Production and Formulations: Fungi

# Chapter 6

# Production of Conidia by the Fungus *Metarhizium anisopliae* Using Solid-State Fermentation

**Octavio Loera-Corral, Javier Porcayo-Loza, Roberto Montesinos-Matias, and Ernesto Favela-Torres**

## Abstract

This chapter describes the production of conidia by *Metarhizium anisopliae* using solid-state fermentation. Before production of conidia, procedures for strains conservation, reactivation, and propagation are essential in order to provide genetic stability of the strains. The strain is conserved in freeze-dried vials and then reactivated through insect inoculation. Rice is used as a substrate for the conidia production in two different bioreactors: plastic bags and tubular bioreactor. The $CO_2$ production in the tubular bioreactors is measured with a respirometer; this system allows calculating indirect growth parameters as lag time ($t_{lag}$) (25–35 h), maximum rate of $CO_2$ production ($rCO_{2\,max}$) (0.5–0.7 mg/gdm h), specific rate of $CO_2$ production ($\mu$) (0.10–0.15 1/h), and final $CO_2$ production ($CO_2$) (100–120 mg/gdm). Conidial yield per gram of dry substrate (gdm) should be above $1 \times 10^9$ conidia/gdm after 10 days of incubation. Germination and viability of conidia obtained after 10 days of incubation should be above 80 % and 75 %, respectively. Bioassays using of *Tenebrio molitor* as a host insect should yield a final mortality above 80 %.

**Key words** Entomopathogenic fungi, *Metarhizium anisopliae*, *Tenebrio molitor*, Solid-state fermentation, Biological control

## 1 Introduction

In the last six decades, chemical pesticides have been the most used tools in insect control or against weeds and plant diseases; however, the continuous accumulation affects the environment and the human health [1]. Alternative programs, such as those based on classical and augmentative biological control and sterile insect techniques are likely to provide effective and sustainable options for the control of native and exotic pests [2]. Once wild fungal isolates have been recovered from agricultural fields, and identified as entomopathogenic strains, feasible conservation methods are essential for long-term studies or industrial productions; additionally, those are reliable methods to preserve genetic stability of the strains [3]. Solid-state fermentation (SSF) is the preferred system

Travis R. Glare and Maria E. Moran-Diez (eds.), *Microbial-Based Biopesticides: Methods and Protocols*, Methods in Molecular Biology, vol. 1477, DOI 10.1007/978-1-4939-6367-6_6, © Springer Science+Business Media New York 2016

to produce conidia from entomopathogenic fungi, mainly using trays of plastic bags containing substrates such as rice or other solid agricultural wastes which sometimes are supplemented or combined in order to achieve higher conidial yields [4]. Moreover, during conidia production the quality of conidial batches should be corroborated; some quality parameters include germination and viability of conidia, since those are related to virulence against insect [5]. This work presented the conidia production of *M. anisopliae* under two different techniques using SSF: plastic bags and tubular bioreactors, as mentioned earlier the first is the most used production method for entomopathogenic fungi; although tubular bioreactors can be monitored online, obtaining process data such as temperature, $CO_2$ production and $O_2$ consumption during the culture. These data have proved useful for comparing diverse treatments such as solid substrates, media supplements, or even strains with different phenotypes [6, 7]. In order to implement similar strategies, including the quality of conidial batches, the methodology is described in this chapter.

## 2 Materials

### 2.1 Organisms

1. The fungus *Metarhizium anisopliae* is stored in freeze-dried vials.
2. The insect *Tenebrio molitor* is kept under laboratory conditions (*see* **Note 1**).

### 2.2 Culture Media

Prepare all your reagents with distilled water and use analytical grade reagents.

1. Tween 80 (0.05 %) for conidia recovery: dissolve 0.05 g of Tween 80 in 100 mL of distilled water in a 250 mL Erlenmeyer flask covered with a cotton plug. Sterilize at 121 °C for 15 min.
2. Sabouraud dextrose agar (SDA) with 0.2 % yeast extract medium (SDAY): dissolve 13 g of SDA and 0.4 g of yeast extract in 200 mL of hot distilled water in a 250 mL Erlenmeyer flask. Add 30 mL of medium to three 125 mL Erlenmeyer flasks. Sterilize the flasks covered with a cotton plug containing 30 mL (three flasks) and 110 mL (one flask) of medium at 121 °C for 15 min. Then, fill five sterile Petri dishes with 20 mL of medium from the Erlenmeyer flask containing 110 mL of sterile medium.
3. SDA with 0.2 % yeast extract and 0.05 % sodium deoxycholate (SDAY-SD): dissolve 6.5 g of SDA, 0.2 g yeast extract, and 0.05 g of sodium deoxycholate in 100 mL of hot distilled water in a 250 mL Erlenmeyer flask. Sterilize the flask covered with a cotton plug at 121 °C for 15 min. Then, fill 5 sterile Petri dishes with 20 mL of sterile medium [3, 5].

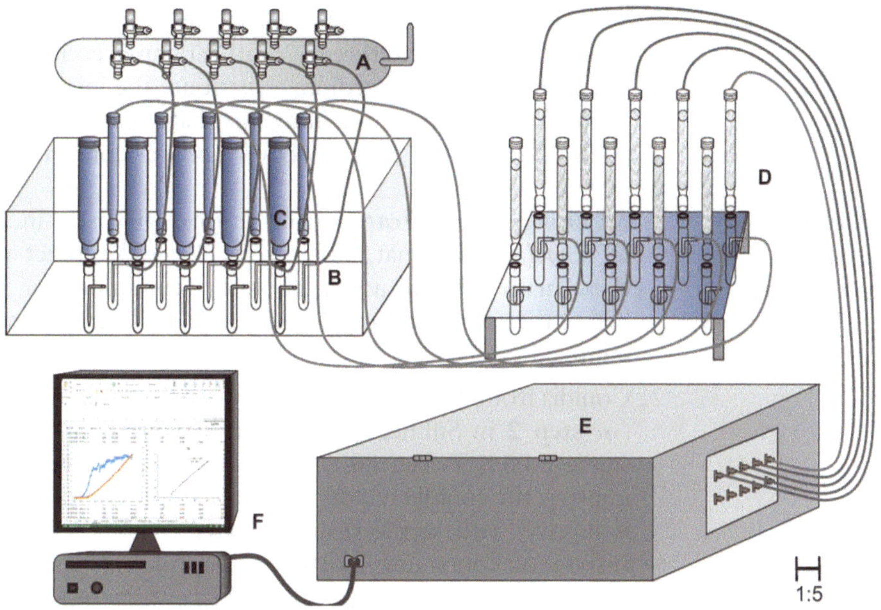


**Fig. 1** Solid-state fermentation and respirometric analysis apparatus. (**a**) Air distributor, (**b**) Water bath, (**c**) Solid-state culture bioreactors, (**d**) Air dryers, (**e**) Respirometer for $CO_2$, $O_2$, and air flow rate measure and (**f**) Computer

#### *2.3 Conidia Production in Plastic Bag*

1. Substrate: pretreated rice (*see* **Note 2**).
2. Twelve plastic bags (30×40 cm).
3. One sterile syringe of 20 mL syringe with needle.
4. One incubation chamber at 28 °C.

#### *2.4 Conidia Production in Tubular Bioreactor*

1. Substrate: pretreated rice (*see* **Note 2**).
2. Tubular bioreactors for conidia production. Twenty tubular glass bioreactors, of which ten have 2.1 cm of internal diameter and 25 cm of height, and the other ten have 4.5 cm of internal diameter and 25 cm of height. Twenty glass humidifiers.
3. One sterile syringe of 20 mL syringe with needle to inoculate the substrate contained in the plastic bags.
4. One air distributor system (Fig. 1).
5. One respirometer.
6. Water bath at 28 °C to incubate the tubular bioreactors.

## 3 Methods

#### *3.1 Metarhizium Anisopliae Propagation*

1. Lyophilized *M. anisopliae* conidia from freeze-dried vials is suspended in 3 mL of sterile 0.05% Tween 80 and homogenized for 10 s in a vortex.

2. With an automatic pipette and a sterile tip add 500 μL of the conidia suspension to three 125 mL Erlenmeyer flasks containing 30 mL of SDAY medium. Distribute the inoculum on the surface of the culture medium. Incubate the inoculated flasks at 28 °C for 10–15 days to obtain a proper conidiation level.

### *3.2 M. Anisopliae Reactivation in Tenebrio Molitor*

1. *M. anisopliae reactivation* is made through insect inoculation in *T. molitor*. For that, 25 larvae of *T. molitor* insect are used; these larvae are kept under laboratory conditions for 1 month before the assay (*see* **Note 1**). All larvae of *T. molitor* are surface sanitized (*see* **Note 3**).
2. Conidia from *M. anisopliae* from one of the three 125 mL flasks (*see* **step 2** in Subheading 3.1) are harvested with 30 mL of sterile 0.05% Tween 80. A sterile magnetic stirrer is used to improve the conidia recovery. After that, the conidia suspension is diluted with sterile distilled water to obtain 30 mL of a suspension containing $1 \times 10^7$ conidia/mL.
3. Each larvae of *T. molitor* is submerged into the conidia suspension for 3 s and the excess of liquid is removed with sterile filter paper.
4. Infected larvae are incubated in the incubation chamber in groups of five at 28 °C for 5–10 days, insects die during this period.
5. Once the insects die, they are separated and incubated alone into a sterile Petri dish placed into the incubation chamber for 10–15 days at room temperature to allow high conidiation level of *M. anisopliae*. Each insect produce above $8 \times 10^7$ conidia/larvae.
6. Conidia samples from the infected insect are recovered with a bacteriological loop and used to inoculate by streaking three Petri dishes (100 × 15 mm) containing 20 mL of SDAY-SD medium each.
7. Cultures of *M. anisopliae* are incubated at 28 °C for 10–15 days.
8. After incubation, conidia reactivated in SDAY-SD medium are recovered with a bacteriological loop and suspended in 5 mL of 0.05% Tween 80. The suspension is standardized at $1 \times 10^4$ conidia/mL.
9. Place 0.5 mL of the standardized suspension (*see* **step 8** in Subheading 3.2) on Petri dishes (100 × 15 mm) containing 20 mL of SDAY medium. Distribute the conidia suspension over the surface of the culture medium with a sterile l-shaped glass tube and incubate at 28 °C.
10. Observe the conidia after 12–16 h of incubation. Count germinated conidia with an optic microscope (40×). A conidium is considered to be germinated if the germination tube is

greater than the diameter of the nongerminated conidia [8]. Conidia germination should be above 80%.

11. Transfer one 2 mm square of medium containing one germinated conidia to a Petri dish with 20 mL of SDAY medium and incubate it at 28 °C for 10–15 days. This procedure allows obtaining monosporic cultures of *M. anisopliae* [9].
12. Recover the monosporic colony with a bacteriological loop and suspend in 5 mL of 0.05% Tween 80. The suspension is then standardized at $1 \times 10^4$ conidia/mL. Place 30 μL of the standardized suspension (300 conidia/Petri dish) on three Petri dishes (100 × 15 mm) containing 20 mL of SDAY-SD medium. Distribute the conidia suspension over the surface of the culture medium with a sterile l-shaped glass tube and incubate at 28 °C for 72 h.
13. Colony forming units (CFU) in each Petri dish are counted. Viability in SDAY-SD medium should be above 75%.

### 3.3 Strain Conservation

1. Add 10 g of skimmed milk in 100 mL of distilled water. Sterilize the suspension at 121 °C for 10 min. Cool the suspension at room temperature.
2. Submerge ten glass ampoules in 2% sodium hypochlorite for 3 h. Then, wash the ampoules three times with distilled water. In the top of the ampoules place a small plug of cotton and sterilize them at 121 °C for 2 h. The obtained monosporic culture (*see* **step 11** in Subheading 3.2) is recovered with a bacteriological loop, suspended in 15 mL of sterile 10% skimmed milk and standardized at a final concentration of $1 \times 10^7$ conidia/mL.
3. One milliliter of the standardized conidia suspension containing $1 \times 10^7$ conidia/mL is placed in each of those ten sterile glass ampoules.
4. A small cotton plug is placed in the top of each glass ampoules.
5. All ampoules are frozen making them rotate in acetone–dry ice mixture.
6. Frozen samples are freeze-dried in a lyophilizer (LABCONCO®) previously stabilized at −50 °C and 10 μm of Hg.
7. After 7 days, three lyophilized ampoules are randomly sampled and evaluated for conidia viability (*see* **step 2** in Subheading 3.8). Conidia viability should be above 70%.

### 3.4 Inocula Production

1. Conidia from a lyophilized ampoule (*see* Subheading 3.3) are suspended in 10 mL of sterile 0.05% Tween 80 and homogenized for 10 s in a vortex.
2. Add 500 μL of suspension to three 125 mL Erlenmeyer flasks containing 30 mL of SDAY medium. Inoculated flasks are incubated at 28 °C for 10–15 days to obtain a proper conidiation level.

### 3.5 Conidia Production in Plastic Bags

1. Pretreated rice (*see* **Note 2**) is used as sole substrate for conidia production.
2. Twenty-five plastic bags containing 200 g of wet rice were sterilized at 121 °C for 20 min. Bags are cooled for 1 h until room temperature.
3. Conidia from three 125 mL Erlenmeyer flasks (*see* **step 2** in Subheading 3.4) are harvested with 30 mL of sterile 0.05 % Tween 80. Conidia are counted with a hemocytometer in an optic microscope (40×); this procedure is made three times (*see* **Note 4**). Erlenmeyer flasks present a conidiation level above $3.5 \times 10^7$ conidia/cm$^2$.
4. The conidia suspension is diluted with sterile distilled water containing 50 ppm chloramphenicol to a final volume of 1.25 L of a standardized inoculum with $2.9 \times 10^6$ conidia/mL.
5. Twenty-milliliter sterile plastic syringes are used to inoculate rice in 23 plastic bags. Each plastic bag is inoculated with 47 mL of the standardized inoculum (*see* **step 4** in Subheading 3.5). Moisture content should be from 40 to 45 % and pH from 5 to 6.5 (*see* **Notes 5** and **6**). Inoculated plastic bags are also used to fill the tubular bioreactors (*see* **step 1** in Subheading 3.6). Two plastic bags with pretreated rice and without inoculum are prepared as control.
6. Plastic bags are kept in a laminar flow hood for 1 h and mixed every 15 min to ensure the homogeneity of the inoculum. After that, they are incubated at 28 °C for 10–15 days into a chamber with water-saturated atmosphere.
7. After 4 days of incubation two plastic bags are sampled every 2 days. Conidia are recovered with 0.05 % Tween 80 and counted with a hemocytometer (*see* **Note 4**). Conidia production after 10–15 days of incubation should be above $1 \times 10^9$ conidia/gdm.

### 3.6 Conidia Production in Tubular Bioreactors

1. Twenty tubular glass bioreactors with two different internal diameters, ten with 2.1 cm of internal diameter (ID) and 25 cm of height, and ten with 4.5 cm of ID and 25 cm of height are used. Place plugs of cotton at the bottom and the top of the bioreactors. Weigh the bioreactors separately and sterilize them at 121 °C for 15 min. After sterilization, take out the cotton in the top and fill the bioreactors with the inoculated substrate (*see* **step 5** in Subheading 3.5). Place the cotton pieces on the top and weigh the bioreactors separately. Each bioreactor is filled with inoculated rice from a plastic bag up to 12 cm of height (~22 gdm and ~72 gdm in tubular bioreactors with ID of 2.1 cm and 4.5 cm, respectively). Each tubular bioreactor is assembled with a glass humidifier and incubated in a water bath at 28 °C with an aeration rate of 0.1 VKgM for 10–15 days (*see* **Note 7**).

2. The output gas of each tubular bioreactor is connected to a respirometer to analyze airflow rate and $CO_2$ concentration (Fig. 1).
3. After 4 days of incubation, two bioreactors are sampled every 2 days. After extraction from the tubular bioreactors, the content of bioreactors is axially divided into three fractions of 4 cm each. Recover the conidia from each fraction with 0.05 % Tween 80 and count the suspension using a hemocytometer (*see* **Note 4**). Conidia production after 10–15 days of incubation should be above $1 \times 10^9$ conidia/gdm.
4. The lag time should be from 25 to 35 h; after that, maxima $CO_2$ production rate (rCO2 max) attains values from 0.5 to 0.7 mg/gdm h. The specific $CO_2$ production rate is from 0.10 to 0.15 1/h and final $CO_2$ production is from 100 to 120 mg/gdm.
5. Moisture content and pH during culture are from 40 to 45 % and 5.0 to 6.5, respectively (*see* **Notes 5** and **6**).

### 3.7 Evaluation of Conidia Quality

1. Conidia quality is evaluated in samples obtained at the end of culture (around 10 days of incubation at 28 °C). Conidia are recovered from samples with 9 mL of 0.05 % Tween 80 per gram of wet matter. Shake the samples for 1 min in a vortex and follow the instructions for each procedure as follows.
2. Conidia germination: Use SDAY medium. Fill five Petri dishes (100 × 15 mm) with 20 mL of sterile medium and inoculate with 0.5 mL of the standardized conidia suspension. Distribute the conidia suspension over the surface of the culture medium with a sterile l-shaped glass tube and incubate at 28 °C. Observe the conidia after 12–16 h of incubation. Count germinated conidia with an optic microscope (40×). A conidium is considered to be germinated if the germination tube is greater than the diameter of the nongerminated conidia [8]. Conidia germination should be above 80 %.
3. Conidia viability: Use SDAY-SD medium. Fill five Petri dishes (100 × 15 mm) with 20 mL of sterile medium and inoculate with 30 μL of standardized conidia suspension at $1 \times 10^4$ conidia/mL (300 conidia/Petri dish). Sterilized l-shaped glass tube is used to distribute the standardized conidia suspension over the surface of the culture medium. Incubate at 28 °C for 72 h. Count the number of colonies (CFU) and reported as a percentage of viability. Conidia viability should be above 75 %.
4. Conidia infectivity: *T. molitor* (*see* **Note 1**) and a standardized conidia suspension containing $1 \times 10^7$ conidia/mL are used to evaluate the conidia infectivity. First, take 150 larvae (50 larvae are used with conidia from plastic bags, 50 larvae with conidia from bioreactors, and 50 larvae as control); sanitize the larvae

surface (*see* **Note 3**). Separate in groups of ten in 50 mL Falcon tubes. Immerse for 5 s each group of larvae in the standardized conidia suspension. Control larvae are submerged in 0.05 % Tween 80. Remove the excess of moisture of larvae with an absorbent paper. Place the inoculated larvae in Petri dishes. Sterile oat-wheat bran (1:1) with moisture content around 60 % (w/v) is used to feed the insect during the bioassay. Incubate the larvae at 28 °C in an incubation chamber with at least 60 % of relative humidity. Register every day the larvae survivors.

## 4 Notes

1. *Tenebrio molitor* are used with a weight between 0.07 and 0.13 g. The insects are kept in a plastic container (22 × 35 × 10 cm) under laboratory conditions (20–25 °C, 40–50 % HR and natural photoperiod) for 1 month before the assay. The insects are fed with a sterile oat–rice 1:1 mixture with moisture content around 60 % (w/v) and a wet cotton plug is used as water provider.
2. Rice used for conidia production is pretreated before use. First, the rice is placed in a big plastic container and it is covered with distilled water (1 L/kg). The wet rice is stirred for 1 min, excess of water is removed and then fresh water is added. This process is made three times. In the last repetition, water is kept for 30 min and then the excess of water is removed. The rice is drained for 20 min; afterward, plastic bags are filled with 200 g of wet rice. The rice in the plastic bags has a moisture content between 30 and 32 % and a pH near to 6 (*see* **Notes 6** and 7).
3. Before inoculation, all insect larvae are submerged in 1 % sodium hypochlorite for 3 s and immediately washed three times with sterile distilled water.
4. Conidia recovery is made homogenizing the material from bags or the bioreactor sections for 1 min. Using a clean spatula, take three material samples (rice and conidia) weighing one gram each in Falcon tubes. Then, add 9 mL of 0.05 % Tween 80 to each tube and homogenize for 1 min in a vortex. Make three dilutions: 1:2, 1:10, 1:20 and count it in hemocytometer. The correct dilution is one that has between 20 and 100 conidia per square of hemocytometer at 40× (*see* ref. 3).
5. pH is evaluated with three samples from the plastic bag or the bioreactor section. Place 1 g of homogenized material (rice and conidia) in a 50 mL Falcon tube. Then, add 9 mL of 0.05 % Tween 80 to each tube and homogenize for 1 min in a vortex. Determine the pH using a calibrated pH meter.

6. Moisture is evaluated with three samples from the plastic bag or the bioreactor section. Place 0.5–1 g of homogenized material (rice and conidia) in a thermobalance with a 90 °C dry program.
7. VKgM = volume of air per mass of initial wet substrate per minute [L/Kg min].

## Acknowledgement

This work was supported by *Universidad Autónoma Metropolitana-Iztapalapa*, the *Secretaría de Ciencia, Tecnología e Innovación* (SECITI), D.F. and the Centro Nacional de Referencia de Control Biológico, Tecomán, Colima, México.

### References

1. Heckel DG (2012) Insecticide resistance after silent spring. Science 337:1612–1614
2. Williams T, Arredondo-Bernal HC, Rodríguez-del-Bosque LA (2013) Biological pest control in México. Annu Rev Entomol 58:19–40
3. Inglis GD, Enkerli J, Goettel MS (2012) Laboratory techniques used for entomopathogenic fungi: hypocreales. In: Lawrence AL (ed) Manual of techniques in invertebrate pathology, 2nd edn. Academic press, United Kingdom, pp 190–192, 215–216
4. Nuñez-Gaona O, Saucedo-Castañeda G, Alatorre-Rosas R, Loera O (2010) Effect of moisture content and inoculum on growth and conidia production by *Beauveria bassiana* on wheat bran. Braz Arch Biol Technol 53:771–777
5. Montesinos-Matías R, Viniegra-González G, Alatorre-Rosas R, Loera O (2011) Relationship between virulence and enzymatic profiles in the cuticle of *Tenebrio molitor* by 2-deoxy-d-glucose-resistant mutants of *Beauveria bassiana* (Bals.) Vuill. World J Microbiol Biotechnol 27:2095–2102
6. Saucedo-Castañeda G, Trejo-Hernández MR, Lonsane BK, Navarro JM, Roussos S (1994) On-line automated monitoring and control systems for CO2 and O2 in aerobic and anaerobic solid-state fermentations. Process Biochem 29:13–24
7. Saucedo-Castañeda G, Favela-Torres E, Viniegra-González G, Torres-Mancera MT, Figueroa-Montero AA, Rosales-Zamora G (2013) Sistema de respirometría con administración remota para el monitoreo en línea de la concentración de CO2 y oxígeno y flujo de los gases de salida en procesos biológicos. Solicitud de patente IMPI, MX/a/2013/004638
8. Safavi SA, Shah FA, Pakdel AK, Rasoulian GR, Bandani AR, Butt TM (2007) Effect of nutrition on growth and virulence off the entomopathogenic fungus *Beauveria bassiana*. FEMS Microbiol Lett 270:116–123
9. Ayala-Zermeño MA, Montesinos-Matías R, Berlanga-Padilla A (2012) Obtención de cultivos monospóricos de hongos entomopatógenos. Centro Nacional de Referencia de Control Biológico, Dirección General de Sanidad Vegetal, SENASICA. Ficha Técnica CB-27, pp. 4

# Chapter 7

# Liquid Culture Production of Fungal Microsclerotia

**Mark A. Jackson and Angela R. Payne**

## Abstract

Fungal microsclerotia ("small" sclerotia) are compact hyphal aggregates, typically 50–600 μm in diameter, that are formed under unfavorable nutritional and/or environmental conditions. These structures are often melanized and desiccated to some degree containing endogenous nutritional reserves for use when favorable conditions return. Many fungi, mostly plant pathogens, produce microsclerotia as a survival structure. Liquid culture methods have been developed for producing microsclerotia of the Ascomycota *Metarhizium spp*, *Colletotrichum truncatum*, *Mycoleptodiscus terrestris*, and *Trichoderma spp.* While these fungi have varying culture conditions that optimize microsclerotia production, all share common nutritional and environmental requirements for microsclerotia formation. Described are the general liquid culture techniques, media components, and harvesting and drying methods necessary to produce stable microsclerotial granules of these fungi.

**Key words** Microsclerotia, Liquid fermentation, Fungal differentiation, Desiccation tolerance, Microbial biocontrol, Biopesticide

## 1 Introduction

Morphologically, sclerotia can be large, well-formed structures with differentiated tissues (true sclerotia) or medium to small aggregates of hyphae (microsclerotia) with no organized structure [1]. The principal biological function for all sclerotial structures is survival during adverse conditions for growth. Fungal sclerotia are well suited as survival structures as they typically survive desiccation, are resistant to oxidative stress and UV radiation, and are known to produce antimicrobial compounds and compounds that deter insect feeding [2]. When conditions amenable to growth return, fungal sclerotia utilize endogenous nutritional reserves to germinate hyphally to infect or colonize host plant tissues or sporogenically to produce conidia capable of infecting host insects or plants. These qualities make microsclerotia well suited for use as granular biopesticides for control of insects, weeds, or plant pathogens.

The development of liquid culture techniques for producing fungal microsclerotia has piqued commercial interest in their use in

Travis R. Glare and Maria E. Moran-Diez (eds.), *Microbial-Based Biopesticides: Methods and Protocols*, Methods in Molecular Biology, vol. 1477, DOI 10.1007/978-1-4939-6367-6_7, © Springer Science+Business Media New York 2016

biological control and provided a method for basic researchers to investigate fungal differentiation in a controlled, homogenous environment [3, 4]. Early studies led to the development of liquid culture methods for producing microsclerotia of the plant pathogens *Colletotrichum truncatum* and *Mycoleptodiscus terrestris* [5, 6]. Both of these fungi were known to colonize their plant hosts and produce microsclerotia in senescing tissues [7]. Recently, we have shown that microsclerotia can be produced in liquid culture under specific nutritional and environmental conditions by two fungal genera that have not been reported to produce sclerotia in nature, *Metarhizium* and *Trichoderma* [8, 9]. In general, these fungi required nutritional environments rich in carbon coupled with environmental conditions that supply adequate quantities of oxygen for differentiation to form microsclerotia. The development of liquid culture methods for the production of stable, effective microsclerotial formulations by various fungi that may or may not produce these structures in nature suggests that nutritional and environmental conditions can be manipulated to induce other potential fungal biocontrol agents to produce microsclerotia. Below, we describe the basic nutritional and environmental conditions required for the differentiation of these fungi to form sclerotia using liquid culture fermentation.

## 2 Materials

All fungal cultures should be maintained as pure cultures without the use of antibacterial supplements. Conidial or hyphal inocula are produced from fungal cultures incubated at room temperature (22 °C) on potato dextrose agar (PDA) plates. Inocula should be from freshly sporulated or actively growing cultures. Distilled or deionized water should be used for all culture media, stock solutions, and inoculum water. It is recommended that a media formulation sheet be used to organize experimental data (Fig. 1).

### 2.1 Culture Maintenance and Growth

1. 10% (w/v) glycerol solution, sterile.
2. PDA (Difco) in Petri plates (100 × 15 mm).
3. Water agar (2% agar) plates (100 × 15 mm).
4. Complete liquid medium for microsclerotia production (all values are per liter deionized water): 75 g Glucose, 15 g acid-hydrolyzed casein (Difco, Casamino acids), 2.0 g $KH_2PO_4$, 0.4 g $CaCl_2·2H_2O$, 0.3 g $MgSO_4·7H_2O$, 50 mg $FeSO_4·7H_2O$, 37 mg $CoCl_2·6H_2O$, 16 mg $MnSO_4·H_2O$, 14 mg $ZnSO_4·7H_2O$, 500 μg each thiamin, riboflavin, pantothenate, niacin, pyridoxamine, thiotic acid, and 50 μg each folic acid, biotin, vitamin $B_{12}$ (*see* **Note 1**).

**MEDIA FORMULATION**

EXPERIMENT # ____________ DATE: ____________

MEDIUM: Basal Medium REPLICATES: 3

pH: uncontrolled ACID: - BASE: -

FLASK: 250ml baffled ORGANISM: ____________

RPM: 300 TEMP: 28oC INCUBATOR: ____________

| Ingredient | 1 | 2 | 3 | 4 | 5 | 6 | 7 | 8 | 9 | 10 | 11 | 12 |
|---|---|---|---|---|---|---|---|---|---|---|---|---|
| Basal Medium(2X)(ml) | 50 | | | | | | | | | | | |
| Glucose(20%)(ml) | 37.5 | | | | | | | | | | | |
| Casamino Acids(g) | 1.5 | | | | | | | | | | | |
| DH2O(ml) | 2.5 | | | | | | | | | | | |
| Inoculum(5E+06) | 10 | | | | | | | | | | | |
| | | | | | | | | | | | | |
| | | | | | | | | | | | | |
| | | | | | | | | | | | | |
| | | | | | | | | | | | | |
| | | | | | | | | | | | | |
| | | | | | | | | | | | | |
| | | | | | | | | | | | | |
| Total Volume | 100 | | | | | | | | | | | |

| BASAL MEDIUM (2X) | 200 mL | 20% GLUCOSE (w/v) |
|---|---|---|
| VITAMIN MIX (50X) | 8 ml | 200 ml |
| $KH_2PO_4$ | 0.8 g | 40 g |
| $MgSO_4$ | 0.12 g | -autoclave separately |
| $CaCl_2$ | 0.16 g | INOCULUM D-$H_2O$ 40 ml |
| $FeSO_4$ | 0.02 g | |
| Zn, Mn, Co (100X stock solutions) | 4 ml each | |
| D-$H_2O$ | 180 ml | |

**Fig. 1** Formulation sheet for construction of the microsclerotia production medium

5. Cobalt chloride stock solution (100×): 1.83 g $CoCl_2{\cdot}6H_2O$ in 500 mL deionized water. Store at 4 °C.
6. Manganese sulfate stock solution (100×): 0.78 g $MnSO_4{\cdot}H_2O$ in 500 mL deionized water. Store at 4 °C.
7. Zinc sulfate stock solution (100×): 0.70 g $ZnSO_4{\cdot}7H_2O$ in 500 mL deionized water. Store at 4 °C.

8. Vitamin stock solution (50×): 25 mg each thiamin, riboflavin, pantothenate, niacin, pyridoxamine, thiotic acid, and 2.5 mg each folic acid, biotin, vitamin $B_{12}$ in 1 L deionized water. Store at 4 °C, light sensitive.
9. Sterile deionized water, 40 mL.
10. Baffled Erlenmeyer flasks with stopper, 250 mL (*see* **Note 2**).
11. Rotary shaker incubator with 2 cm or greater shaker orbit and refrigeration (*see* **Note 3**).

### 2.2 Microsclerotia Harvesting and Drying

1. Laboratory sieve screen: 120 mesh, 20 cm diameter.
2. Diatomaceous earth (DE) filter aid (*see* **Note 4**).
3. Buchner funnel (128 mm ID), 1 L vacuum flask, Whatman # 1 filters, 125 mm.
4. Drying chamber (Fig. 2) and trays (*see* **Note 5**).
5. Moisture analyzer (*see* **Note 6**)

### 2.3 Measurements

1. Light microscope.
2. Stereomicroscope.
3. Hemacytometer.

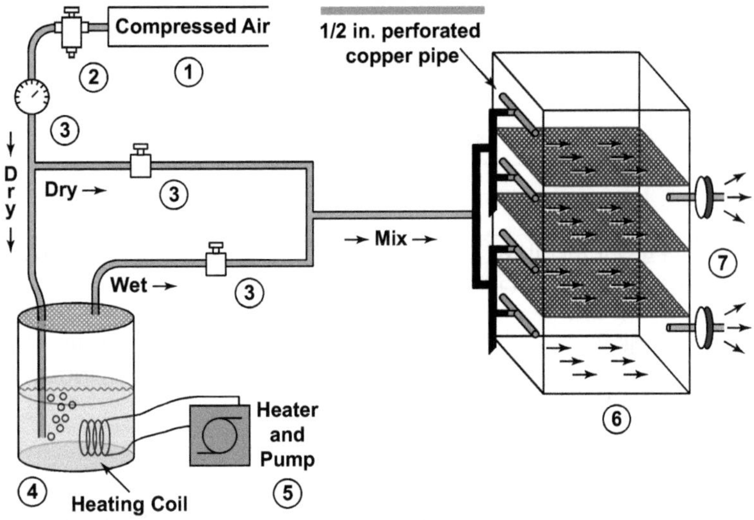


**Fig. 2** Schematic of controlled humidity air-drying chamber. Low RH compressed air (1) was used as the air source with air pressure (2) regulated. Airflow was adjusted with valves (3) that controlled the volume of air and the proportion of wet and dry air delivered to the drying chamber. Moist air was produced by bubbling air through a water bath (4) in which water temperature could be adjusted with a heater (5) and heating coil. The air manifolds produced a stream of air over samples on the shelves of the drying chamber (6). The drying chamber was fitted with air vents (7) that could be filtered or open depending on the samples being dried

4. Wide-bore 1 mL and 100 μL plastic pipet tips, modified (*see* **Note 7**).
5. Wide-bore pipet, 5 and 10 mL (Fisher Scientific, #13-678-35A and 35B).

## 3 Methods

All culture maintenance and growth processes for fungal inoculum and for microsclerotia production were conducted using sterile technique to insure culture purity. A laminar flow or bio-containment hood is recommended. Microsclerotia harvesting and drying protocols were conducted at the bench or in a bio-containment hood and were aseptic but not sterile technique (*see* **Note 8**).

### 3.1 Stock Cultures and Inoculum Development

1. Grow pure fungal cultures on PDA for 2–3 weeks at room temperature. Preserve stock cultures by cutting sporulated or hyphal fungal cultures into 1–2 mm agar pieces, suspend 4–5 pieces in 1 mL of sterile 10% glycerol in a 2 mL cryovial (Corning, #430659) or sterile test tube, and store at −80 °C (*see* **Note 9**).
2. Inoculate PDA plates with agar pieces from stock cultures and incubate for 2–3 weeks to produce conidia or hyphae for inoculating microsclerotia production media (*see* **Note 10**).

### 3.2 Shake Flask Microsclerotia Production

1. Prepare a 2× concentrate of the basal medium, trace metals, and vitamins in an appropriate volume of deionized water (*see* **Note 11**). Add 50 mL of the 2× concentrated medium to the 250 mL baffled flask along with the nitrogen source and 2.5 mL makeup water (Fig. 1). Prepare flask with closure for autoclaving.
2. Prepare a 20% glucose stock solution by adding 100 mL of deionized water and a magnetic stir bar to a 250 mL graduated cylinder. Place graduated cylinder on a stir plate and add 40 g of glucose. Mix until dissolved. QS with deionized water to 200 mL and transfer glucose solution to appropriate bottle for autoclaving.
3. Prepare one bottle of deionized water (40 mL) for rinsing spores from PDA plates for use as inoculum (*see* **Note 12**).
4. Autoclave the baffled flasks containing the basal medium, nitrogen source, and makeup water and the bottles containing the 20% glucose solution and inoculum water. Once autoclaved, cool the sterilized solutions to incubation temperature (28 °C).
5. In a bio-containment or laminar flow hood, add 37.5 mL of 20% glucose to each baffled flask to achieve a final glucose concentration of 7.5% (w/v) using sterile technique.

6. In a bio-containment or laminar flow hood or using aseptic technique, obtain a conidial suspension from 2- to 3-week-old sporulated fungal cultures by adding 10 mL of sterile deionized water (*see* **Note 13**). Scrape the plate surface with a sterile loop or pipet to obtain a conidial suspension. Microscopically measure the conidial concentration in the suspension from the agar plate and in the inoculum water using a hemacytometer. Add the spore suspension from the sporulated culture to the inoculum water to achieve a final concentration of $5 \times 10^7$ conidia $mL^{-1}$.
7. Inoculate the microsclerotia production medium (90 mL volume in 250 mL baffled Erlenmeyer flask) with 10 mL of the conidial suspension to produce a culture volume of 100 mL with a final conidial concentration in the microsclerotia production medium of $5 \times 10^6$ conidia $mL^{-1}$ (Fig. 1). If the fungus is non-sporulating, such as *Mycoleptodiscus terrestris*, cut colonized PDA plate into 1–2 mm agar pieces using a sterile scalpel and inoculate the microsclerotia production medium with ¼ of the colonized plate taking the agar pieces from the center of the plate to avoid contaminants. If using an agar inoculum, add 10 mL makeup water to obtain a culture volume of 100 mL.
8. Incubate the microsclerotia production flasks in a rotary shaker incubator at 300 rpm and 28 °C. Examine shake flasks frequently during culture growth and keep fungal biomass off the flask walls and in the liquid medium (*see* **Note 14**).
9. Incubate cultures for 7–10 days. Microscopically evaluate shake flask cultures periodically (every 2–3 days) to observe culture morphology and to measure microsclerotia formation and biomass yields (*see* **Note 15**). Do not remove flasks from shaker incubator for extended periods of time to avoid exhausting oxygen levels.

### 3.3 Microsclerotia Counting, Harvesting, and Drying

1. Use a 5 mL wide-bore pipet to obtain a 1 mL whole-culture sample from the shake flask for determining microsclerotia concentration. Using the 1 mL whole-culture sample, make two 1:10 serial dilutions to obtain a 1:100 dilution. Due to the viscosity of the whole culture, the repeated dilution of 1 mL whole culture in 9 mL deionized is advised using 1 mL wide-bore plastic pipet tips. Using a 100 μL wide-bore plastic pipet tip, place 100 μL of the 1:100 dilution on a glass microscope slide and cover with an extra large 24 × 50 mm cover slip. Count all compact microsclerotia under the cover slip using a light microscope at low power (4×) (*see* **Note 16**) and calculate their concentration per milliliter.
2. Harvest microsclerotia once they have produced well-formed compact hyphal aggregates. The formation of compact

microsclerotia typically takes 3–4-day incubation with melanization of microsclerotia occurring 4–8 days post-inoculation (Fig. 3), depending on the fungus being cultured.

3. Measure the volume of the whole culture or the whole culture used for obtaining sieved microsclerotia and add 2.5 % (w/v) diatomaceous earth (DE). Mix thoroughly.

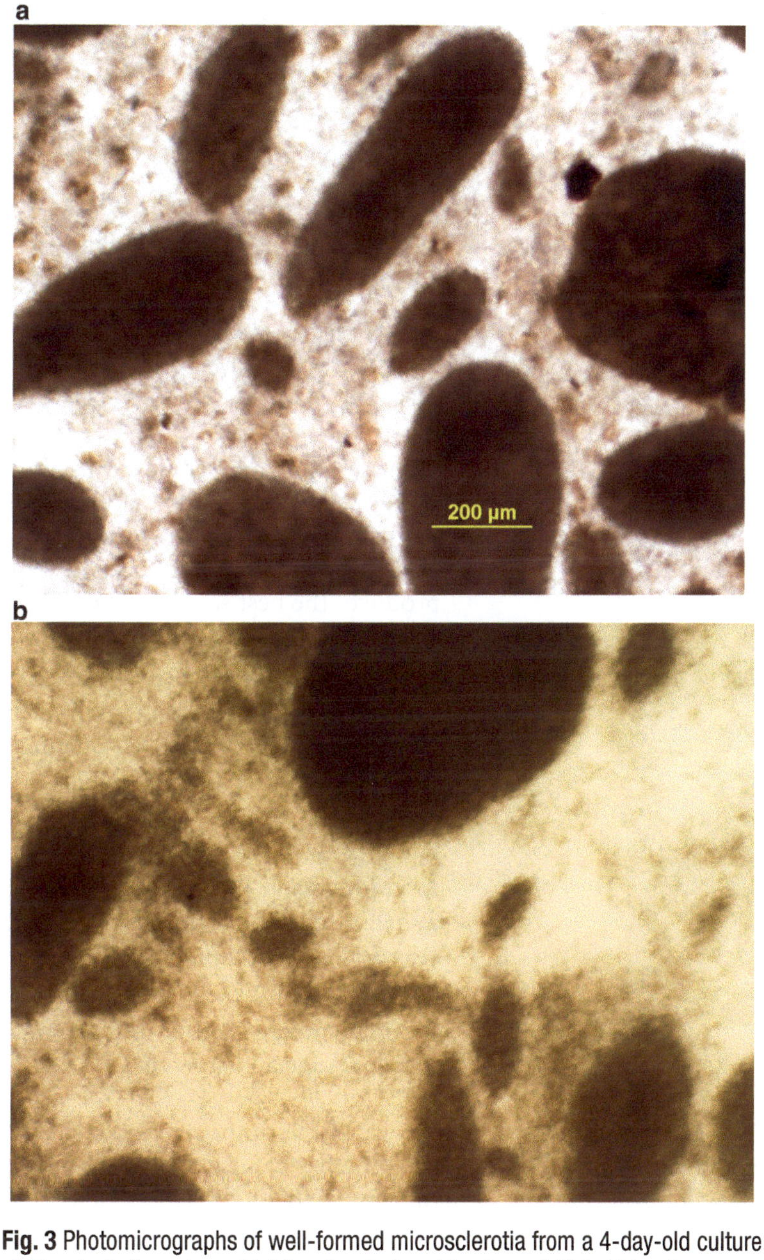


**Fig. 3** Photomicrographs of well-formed microsclerotia from a 4-day-old culture of *Mycoleptodiscus terrestris* (**a**) and an 8-day-old culture of *Metarhizium brunneum* (**b**). The onset of melanization varies with the fungus being cultured

**Fig. 4** Microsclerotia of *Mycoleptodiscus terrestris* collected on a 120 mesh sieve screen. Hyphae have been washed away from microsclerotia using deionized water

4. Microsclerotia may be sieved using an 80 or 120 mesh screen (Fig. 4). Pour the whole culture onto the sieve screen and rinse with water until microsclerotia are the main fungal biomass remaining on the screen. Use the screen (80 or 120 mesh) that produces the best suspension of microsclerotia without hyphae (*see* **Note 17**).
5. Place a #1 Whatman filter in the Buchner funnel, wet the filter with deionized water, and apply vacuum. Pour the microsclerotia-DE mixture into the Buchner funnel and remove the spent media (*see* **Note 18**).
6. When de-watering is complete, remove the filter cake (should be ~ 75 % moisture) and crumble by hand or by pulsing in a food processor. Lay a thin layer (~0.5 cm) of the filter cake in a shallow pan or on aluminum foil (Fig. 5). Air-dry the microsclerotia-DE formulation overnight in a fume hood, biocontainment hood, or drying chamber (*see* **Note 19**).
7. When the microsclerotia-DE formulation is less than 4 % moisture, package and store at 4 °C (*see* **Note 20**).

#### *3.4 Microsclerotia Evaluation*

1. Weigh 25 mg of the microsclerotia-DE formulation and sprinkle onto water agar plates. Incubate at 28 °C for 24 h and measure viability by using a stereomicroscope to evaluate 100 individual microsclerotial granules for hyphal germination into the water agar. Calculate percent viability for microsclerotial granules.

**Fig. 5** Microsclerotial granules of *Metarhizium brunneum* being air-dried in our drying chamber by spreading a thin layer of the DE-microsclerotia mixture on aluminum foil and allowing air to flow over the microsclerotial granules

2. Incubate microsclerotia-DE granules on water agar plates for an additional 7 days at 28 °C to maximize conidia production (Fig. 6). Determine conidia production by microsclerotia-DE granules by rinsing plates with 10 mL of deionized water and scraping the surface of the plate with a plastic loop to dislodge conidia. Collect and measure the volume of the conidial suspension with a 10 mL wide-bore pipet and determine the conidia concentration microscopically with a hemacytometer. The conidia concentration times the liquid volume collected, divided by 0.025, equals conidia produced per gram dried microsclerotia-DE formulation.

## 4 Notes

1. This liquid medium utilizes a high carbon concentration (75 g glucose $L^{-1}$) and low level of nitrogen (15 g acid hydrolyzed casein $L^{-1}$) to bring about the rapid differentiation of the fungal culture to form microsclerotia. Higher levels of nitrogen delay microsclerotia formation but yield higher biomass concentrations. The use of a medium formulation sheet (Fig. 1) aids in construction of a medium. Please note that the use of stock solutions of the trace metals and vitamins reduces repetitive micro-measurements of these nutrients.
2. Baffled flasks coupled with agitation rates of 300 rpm (2 cm or greater shaker orbit) are needed to produce the high-dissolved oxygen levels necessary to support microsclerotia formation and melanization.

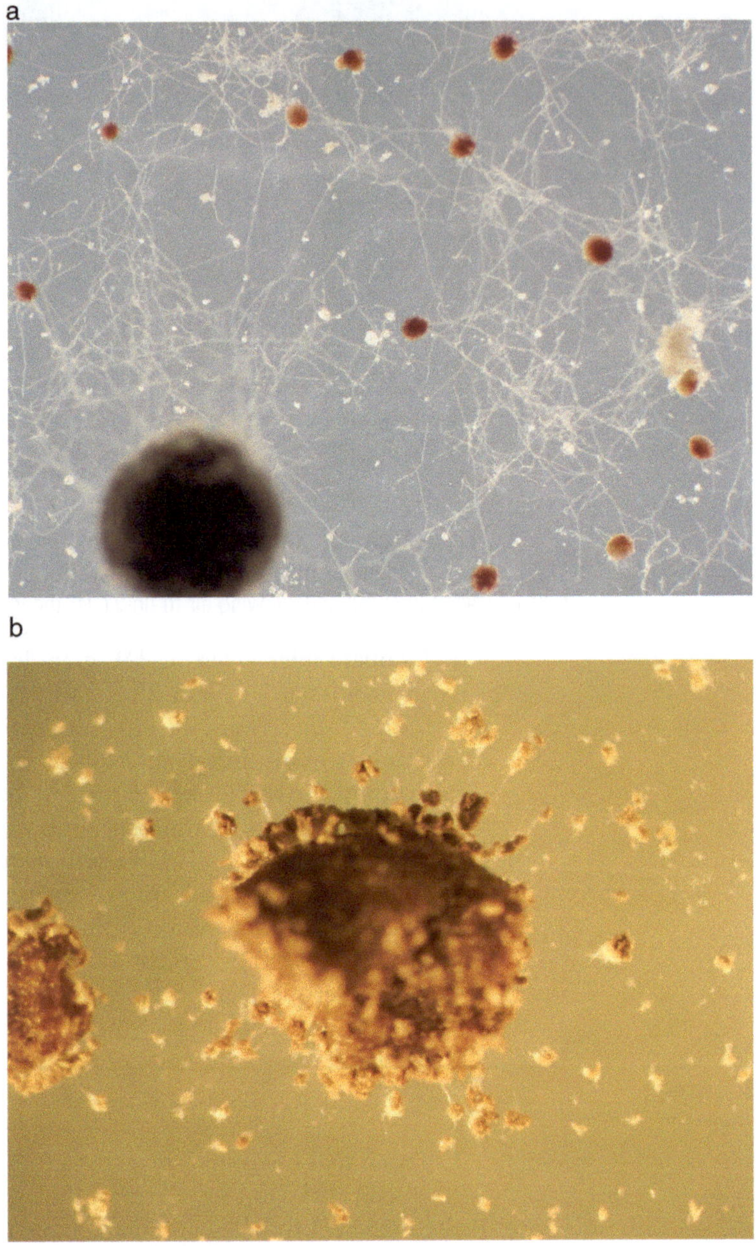


**Fig. 6** Conidia production by microsclerotia-DE granules of *Mycoleptodiscus terrestris* (**a**) and *Metarhizium brunneum* (**b**) on water agar after 8-day incubation at 28 °C. Note satellite conidia production from hyphal extensions of the microsclerotial granules of both fungi

3. Shaker incubators with refrigeration are required to maintain culture temperatures at 28 °C or below.
4. Diatomaceous earth is a non-compressible, inert filter aid required to dewater whole cultures and keep microsclerotia

separated during drying. Without a filter aid like diatomaceous earth, the sclerotia will fuse together when dried producing a dried material that cannot be granulated without fracturing microsclerotia. We use the product (Hyflo®, Celite Corp., Lompoc, CA, USA).

5. Microsclerotia formulations can be dried in a bio-containment, laminar flow, or fume hood. We use a drying chamber where the RH of the drying air can be controlled to dry our MS formulations (*see* ref. [10]). The air supply is compressed air (very low RH) that can be moderated with moist air to produce drying air with an appropriate RH.
6. We use a moisture analyzer (Mark II, Denver Instruments, Arvada, CO, USA) to obtain real-time moisture content measurements. This allows the formulations to be immediately vacuum sealed when there is less than 4% moisture. If a moisture analyzer is not available, package when the formulation appears to be dry and measure moisture using any standard gravimetric method. If later analysis shows that the moisture content is too high, open the vacuum-sealed package and continue air-drying the microsclerotia formulation until the appropriate dryness is obtained.
7. Wide-bore plastic 1 mL and 100 μL pipet tips can be produced by chopping off 1 cm of the tip with a razor blade. Wide-bore glass pipets are needed to obtain larger homogenous microsclerotia samples as microsclerotia can range from 50 to 600 μm in diameter and can clog a normal-bore pipet.
8. The purity of the fungal culture is required during culture maintenance and growth for formation of microsclerotia. The drying and formulation processes should be aseptic but not necessarily sterile technique. Contamination at the formulation and drying steps is insignificant if the formulations are properly dried in a reasonable period of time (24 h) to less than 4% moisture.
9. Each stock culture cryovial can be used to inoculate 5–10 PDA Petri plates. Spread plate 100 μL of the fungal stock culture onto each individual PDA plate.
10. Each fungus can potentially require a different length of incubation time to produce an appropriate conidial or hyphal inoculum on the PDA agar plate. For the *Metarhizium*, *Colletotrichum*, *Mycoleptodiscus*, and *Trichoderma* strains we have worked with, 2–3-week incubation at room temperature is adequate. To ensure that we have fungal inoculum available for liquid culture experiments, we inoculate PDA plates with stock cultures weekly.
11. Concentrated media solutions (basal salts, trace metals, vitamins, glucose) are required in order to obtain the correct "final"

nutrient concentrations in the microsclerotia production medium and to reduce measurement requirements. See media formulation sheet for microsclerotia production medium as an example (Fig. 1).

12. The volume of sterile water required for inoculum depends on the number of flasks to be inoculated and the volume of inoculum. We use a 10% inoculum (10 mL inoculum in 100 mL culture) and calculate the needed spore concentration in the inoculum as 1 log higher than the desired final spore concentration in the culture medium.
13. If the hydrophobicity of the spores prohibits their suspension in water, use an appropriate surfactant such as a sterile aqueous solution of 0.04% Tween 80.
14. The fungal ring on the flask wall can generally be removed by taking the flask from the shaker and vigorously shaking until the biomass on the flask wall is back in the liquid. Adherence to the flask wall is most severe in the first 2 days of growth. If the biomass cannot be removed from the wall by shaking, take the flask to bio-containment hood, remove stopper, and remove with a sterile pipet or inoculating loop. Always return flask to shaker as soon as possible to maintain adequate dissolved oxygen levels in the culture.
15. Using the microsclerotia production medium described herein, the formation of microsclerotia should begin within 2–4-day incubation. After 7–10-day incubation, the microsclerotia should be melanized. Microscopic observation for microsclerotia formation and changes in culture color should be conducted every 2–3 days during culture growth.
16. Only well-formed hyphal aggregates with smooth edges in the 50–600 μm size range are considered microsclerotia. Microsclerotia can be round or oblong. Count all the microsclerotia under the cover slip.
17. When rinsing hyphae away from microsclerotia, use a hose attached to the water faucet. Focus a strong stream of water on the microsclerotia until most of the hyphae are removed. If the microsclerotia are melanized, the biomass will become darker as the hyphae are removed.
18. When removing spent media, be sure that the filter remains under vacuum by pressing the edges of the filter cake or pressing the filter cake together where cracks form in the filter cake. A vacuum must be maintained to ensure proper removal of the spent media.
19. Mix MS-DE formulation periodically with a spatula to ensure the rapid, even drying of the formulation. It is important to dry the formulation under conditions where a constant flow of

air removes moisture. The drying chamber (Jackson and Payne) described in Fig. 2 allows control of the RH of the drying air and aids in standardizing the drying time and final moisture content of the formulation.

20. We typically store the dried microsclerotia-DE formulations in polyethylene bags under vacuum at 4 °C although any sealed container will be adequate if the final moisture content is 4% or lower.

## References


1. Chet I, Henis Y (1975) Sclerotial morphogenesis in fungi. Annu Rev Phytopathol 13: 169–192
2. Coley-Smith JR, Cooke RC (1971) Survival and germination of fungal sclerotia. Annu Rev Phytopathol 9:65–92
3. Jackson MA, Schisler DA (2002) Selecting fungal biocontrol agents amenable to production by liquid culture fermentation. In: Proc 7th Biocontrol Working Group Meeting, Influence of A-Biotic and Biotic Factors on Biocontrol Agents, IOBC WPRS Bulletin pp. 387–391
4. Song Z, Yin Y, Jiang S, Liu J, Chen H, Wang Z (2013) Comparative transcriptome analysis of microsclerotia development in *Nomuraea rileyi*. BMC Genomics 14:411
5. Jackson MA, Schisler DA (1994) Liquid culture production of microsclerotia of *Colletotrichum truncatum* for use as bioherbicidal propagules. Mycol Res 9:879–884
6. Shearer JF, Jackson MA (2006) Liquid culture production of microsclerotia of *Mycoleptodiscus terrestris*: a potential biological control agent for the management of hydrilla. Biol Control 38:298–306
7. Buchwaldt L, Morrall RAA, Chongo G, Bernier CC (1996) Windborne dispersal of *Colletotrichum truncatum* and survival in infested lentil debris. Phytopathology 86: 1193–1198
8. Jackson MA, Jaronski ST (2009) Production of microsclerotia of the fungal entomopathogen Metarhizium anisopliae and their potential for use as a biocontrol agent for soil-inhabiting insects. Mycol Res 113:842–850
9. Kobori NN, Mascarin GM, Jackson MA, Schisler DA (2015) Liquid culture production of microsclerotia and submerged conidia by *Trichoderma harzianum* active against damping-off disease caused by *Rhizoctonia solani*. Fungal Biol 119:179–190
10. Jackson MA, Payne AR (2007) Evaluation of desiccation tolerance of blastosporesof *Paecilomyces fumosoroseus* (Dueteromycotina: Hyphomycetes) using a lab-scale, air-drying chamber with controlled humidity. Biocontrol Sci Technol 17:709–719

# Chapter 8

# Isolation and Assessment of Stability of Six Formulations of Entomopathogenic *Beauveria bassiana*

**Lizzy A. Mwamburi**

## Abstract

*Beauveria bassiana* is the most widely studied and exploited entomopathogen. The development of a suitable formulation for *B. bassiana* is a critical component in aiding the entomopathogen germinate and infect the host. In addition to being economical to produce, having high residual activity, it is also important that the formulation is easy to handle, stable during storage, and convenient to mix and apply and be consistently effective in controlling the target pest.

In this chapter we describe preparation of experimental formulations of conidia of *B. bassiana*. The formulations are prepared with barley, rice, wheat bran, clay, kaolin, and peat. The protocol for assessing the stability of the formulations of *B. bassiana* is also described.

**Key words** *Beauveria bassiana*, Isolation, Formulation, Barley, Rice, Wheat bran, Clay, Kaolin, Stability

## 1 Introduction

Entomopathogenic microbes can serve as alternatives to broad-spectrum chemical insecticides. Numerous advantages including safety for humans and other non-target organisms, reduction of pesticide residues in food, preservation of other natural enemies, and increased biodiversity in managed ecosystems can be found in the utilization of entomopathogens, in addition to efficacy.

Fungal pathogens possess a purely contact mode of action. Infectious propagules must be inoculated onto the target pest or onto substrates in the habitat from which secondary inoculation can be effected via pest movement or feeding [1]. The fungi may be applied directly to the insect as wettable powders, emulsions, or dusts, with conventional equipment used for the application of synthetic chemical insecticides, amended into baits or traps, or added to soil [2].

Travis R. Glare and Maria E. Moran-Diez (eds.), *Microbial-Based Biopesticides: Methods and Protocols*, Methods in Molecular Biology, vol. 1477, DOI 10.1007/978-1-4939-6367-6_8, © Springer Science+Business Media New York 2016

The development of a suitable formulation is a critical component in helping a biological control agent to germinate and infect the host [3]. Additionally, biopesticides must be economical to produce, and have high residual activity. It is also important that they are easy to handle, stable during storage, convenient to mix and apply, and consistently effective in controlling the target pest [3].

*Beauveria bassiana* (Balsamo) Vuillemin (Hyphomycetes) is the most extensively studied and exploited entomopathogen [4, 5]. It is a ubiquitous fungus occurring naturally in many areas of the world [2, 6, 7].

The three developmental stages of *B. bassiana* are conidia, blastospores, and mycelia [2]. Ingredients for the formulation of *B. bassiana* selected should improve spray coverage, including microsite targeting, and rain fastness; increase safety (e.g., reduce dust inhalation, eye irritation); improve and simplify handling; improve storage stability (especially at moderate to high temperatures); improve field stability (especially under ultraviolet radiation); and improve efficacy (especially reduce ambient temperature requirements) [2, 8].

This chapter describes the isolation of *B. bassiana* from soil using the insect bait method [9] and mass production using a diphasic system of six formulations that are cost effective. A method of assessing the stability of the different formulations is also described.

## 2 Materials

### 2.1 *Galleria* Diet

1. Weigh 307 g of maize meal, 225 g honey, 45 g beeswax, and 90 g yeast.
2. Boil the beeswax until melted and then mix with the honey.
3. Mix the maize meal and yeast and then pour into the melted bees wax and honey mixture.
4. Stir the mixture and cook on a hot plate until firm and evenly mixed. Place the mixture in a bowl with a perforated lid and leave to cool overnight.

### 2.2 Quarter-Strength Potato Dextrose Agar

1. Prepare quarter-strength potato dextrose agar (PDA) by adding 9 g of dehydrated PDA to 1 L of distilled water and stir to obtain a uniform suspension.
2. Autoclave the suspension at 120 °C at 15 Psi for 15 min and then allow the medium to cool, to about 30–40 °C in lamina flow airflow cabinet.
3. Add antibiotics (streptomycin sulfate 100 mg/L and penicillin 62.8 mg/L) to the cooled medium and mix using a magnet stirrer to make the medium semi-selective, i.e., to inhibit bacterial growth.

4. Pour PDA media into 9 cm Petri dishes and allow to solidify before use.

### 2.3 Agar slants

1. Prepare agar slant by dispersing 10 mL of autoclaved molten PDA into tilted, screw-capped universal bottles

## 3 Methods

### 3.1 Isolation of *Beauveria bassiana* from the Soil

1. Isolate the fungus from the soil using the waxy moth (*Galleria mellonella*) insect bait method [9].
2. Maintain the colony of *G. mellonella* in the laboratory at room temperatures (20 ± 3 °C) and by feeding on *Galleria* diet.
3. Introduce the *G. mellonella* larvae into the *Galleria* meal and allow the larvae to grow to maturity.
4. Place moist soil samples (w/v 20/80) in Petri dishes and place ten medium-sized larvae into the soil. Turn the dishes regularly in the beginning of baiting period (first week) to make the bait insect larvae penetrate the soil as much as possible while they are still vigorous.
5. Leave the larvae in the soil until the fungus growth is observed on their body.
6. Surface sterilize whole infected larvae that show hyphal growth on their bodies using 70% ethyl alcohol for 3 min to prevent external saprophytic fungi from growing on the dead cadaver.
7. Isolate the fungus from the infected *G. mellonella* moth by scrapping using sterile scalpel into sterile filter paper and transfer the scraps onto the surface of solid quarter-strength PDA in Petri dishes using sterile forceps.
8. Place the fungal scraps in Petri dishes, seal the plates with parafilm to avoid contamination, and then incubate at room temperature (20 °C).
9. Observe the plates after 24 h for the presence of fungal growth and confirm the grown culture by microscopic observation.
10. Subculture the fungus on fresh PDA plates after 4 days, until a pure culture of the fungus is obtained.
11. Place the larvae in PDA medium and incubate at room temperature until adequate growth of the fungus is observed. Transfer the fungus to fresh PDA medium and incubate for 7 days under the same conditions.
12. Code the fungal isolates according to different places from where the soils were collected.

### 3.2 Single Conidium Suspension

1. Transfer conidial mass aseptically to sterile universal bottles containing sterile distilled water by use of sterile wire loop.
2. Streak a loopful onto the surface of the fresh Sabouraud dextrose agar (SDA) media, incubate at 20 °C for 24 h, and then observe under a dissecting microscope for conidia germination.
3. Mark the germinating conidium with a circle on the reverse of the plate using a marker pen, and carefully cut out the marked media portion using a sterile blade transfer to fresh SDA plates using a sterile inoculating needle.
4. Examine the fresh inoculated plates under a dissecting microscope to confirm that only a single germinating conidium was transferred and incubate at 20 °C for 24 h.

### 3.3 Beauveria bassiana Identification

1. *B. bassiana* isolates are identified macroscopically using cultural and microscopically hyphal characteristics.

### 3.4 Culture Maintenance

1. Subculture the single-spore cultures on several plates of SDA media and incubate for 10 days.
2. Maintain the stock cultures for each isolate on SDA slants in sterile universal bottles in refrigerators at 4–5 °C.
3. Using a 2 mm sterile cork borer, cut out agar plugs from the leading margins of fungal cultures and incubate on fresh PDA at room temperature for 7 days.
4. Cover the pure fungal cultures with parafilm before refrigeration.
5. Check the cultures regularly for contamination.

### 3.5 Cultivation of Beauveria bassiana

1. Using a sterile wire loop, transfer spores from the agar slants onto the SDA plates in 9 cm Petri dishes and incubate in complete darkness for 21 days.
2. Inoculate conidia of sub-cultured *B. bassiana* isolates into 50 ml of Sabouraud maltose yeast (SMY) liquid broth in a 250 ml Erlenmeyer flask and incubate aerate vigorously (200 rpm) on a rotary shaker for 2–3 days at 25 ± 2 °C.
3. Approximately 1 ml should be taken out of the flasks everyday to check the developmental stage and also to determine the concentration of the spores using a Neubauer hemocytometer.
4. Homogenize the mycelium in a blender and inoculate in the same liquid medium.
5. Place 10 ml of the culture onto plates containing SMY solid medium and incubate for 10–15 days at 25 °C.

6. Suspend the conidia that develop in the flask in 15 ml of a solution containing 0.2% Tween 20 and 0.89% NaCl.
7. Harvest the conidia by filtration of the conidial suspension using a Buchner-type funnel through a two-layered filter paper (90 mm) to remove mycelial fragments and aggregated conidia and drying the resulting mycelial mat.
8. Determine the spore concentration using a Neubauer hemocytometer under a compound microscope (40×). The final working concentration is obtained by using the formula N/V×D, where $N$ = number of conidia, $V$ = volume of the chamber, and $D$ = dilution factor.

### 3.6 Barley/Rice/Wheat Bran Substrate Formulation

1. Weigh 200 g of the substrate and wash three to four times with sterile distilled water.
2. Pre-cook the substrate by soaking it in 90–100 °C water for 15 min.
3. Place the substrate in polythene bags and autoclave for 60 min at 121 °C.
4. Allow the substrate to cool to about 40–45 °C.
5. Make a small opening in one corner of the bag using sterile scissor.
6. Mix the conidia/spore, mycelia, and conido-mycelia mixture with 1000 ml sterile distilled water.
7. Inoculate the substrate bags with 100 ml the conidial suspension, then seal, and incubate at room temperature for 12 h.
8. Shake the bags manually and incubate for 1 week.
9. After a week, shake the bags slightly to enhance aeration and incubate for another 2 weeks.
10. Transfer the substrate containing conidia into plastic basins and allow it to dry in desiccators by using silica gel for 12 h.
11. At the end of the fermentation, the particulate substrate is unloaded from the bags to a high-shear blender to break up clumps.
12. The materials are then passed through a sieve and any clumps larger than 1.0 mm should be removed.
13. Store the formulations in sealed paper bags in the refrigerator at 4 °C for up to 6 months without the viability being affected before being used in the field.
14. In order to have propagules that are stable, the moisture content should be maintained at 10–30%.
15. To increase the yield of fungal propagules and colonization of particles, the pH of the substrate should be 4–7.

### 3.7 Clay/Kaolin/Peat Formulations of Beauveria bassiana

1. Prepare formulation based on Burges (1998) method.
2. Mix 1 % skimmed milk, 2 % glycerol (*see* **Note 1**), 4 % canola oil (*see* **Note 2**), and 5 % clay (kaolin or peat) (*see* **Note 3**).
3. For each treatment, 1000 ml sterile distilled water containing the required concentration of conidia is added to the final formulation (*see* **Note 4**).

### 3.8 Determination of Stability of Formulations

1. To determine stability of formulations, samples of 10 g of solid substrates are taken at intervals of 4 weeks to 6 months.
2. Prepare serial dilutions of up to $10^{-6}$ of the substrate using sterile distilled water, out of which 1 ml sample is plated onto malt extract agar (MEA) and yeast tryptone agar (YTA) plates, and incubate at 25 °C.
3. Determine the number of colony-forming units (*see* **Note 5**).

## 4 Notes

1. Glycerol is included due to its role as nutrient as humectant, nutrient, and adhesive; whereas skimmed milk acts as nutrient and humectant (*see* ref. 10).
2. Oil is used because it is an excellent adhesive, promoting contact between the active ingredient (the conidia) and the lipophilic insect cuticle while also increasing the conidia's rain fastness on the waxy leaf surface of treated host plants (*see* ref. 10).
3. Clay (kaolin or peat) is added to protect conidia against UV light (*see* ref. 11).
4. Formulations should contain conidia of *B. bassiana* at a concentration of $2.3 \times 10^7$ spores per milliliter or at least $5 \times 10^8$ spores per grams.
5. Colonies may be counted using a colony counter. Plates with more than 300 or less than 30 colonies are not counted. In the former case, the colonies run together, and in the latter, there are too few to allow statistically accurate counts. Once you count the colonies, multiply by the appropriate dilution factor to determine the number of CFU/mL in the original sample.

### References

1. Faria M, Wraight SP (2001) Biological control of *Bemisia tabaci* with fungi. Crop Prot 20:767–778
2. Feng MG, Poprawski TJ, Khachatourians GC (1994) Production, formulation and application of the entomopathogenic fungus *Beauveria bassiana* for insect control: current status. Biocontrol Sci Technol 4:3–34
3. Brar SK, Verma M, Tyagi RD, Valero JR (2006) Recent advances in downstream processing and formulations of *Bacillus thuringiensis* based biopesticides. Process Biochem 41:323–342
4. Glare TR, Milner RJ (1991) Ecology of entomopathogenic fungi. In: Arora DK, Ajello L, Mukerji KG (eds) Handbook of applied mycol-

ogy, vol 2, Humans, animals and insects. Marcel Dekker Inc., New York, pp 547–612

5. Hajek AE, St. Leger RJ (1994) Interactions between fungal pathogens and insect hosts. Annu Rev Entomol 39:293–322
6. Roberts DW, LeBrun RA, Semel M (1981) Control of the Colorado potato beetle with fungi. In: Lashomb JH, Casagrande R (eds) Advances in potato pest management. Hutchinson Ross Pub, Stroudsburg, pp 119–137
7. Humber RA (1996) Fungal pathogens of the chrysomelidae and prospects for their use in biological control. In: Jolivet PHA, Cox ML (eds) Chrysomelidae biology, vol 2, Ecological Studies. SPB Academic Publishing, Amsterdam, pp 93–115
8. Wraight SP, Carruthers RI (1999) Production, delivery, and use of mycoinsecticides for control of insect pests on field crops. In: Hall FR, Menn JJ (eds) Biopesticides, use and delivery. Humana Press, Inc., Totowa, NJ, pp 233–270
9. Goettel MS, Inglis GD, Wraight SP (2000) Fungi. In: Lacey LA, Kaya HK (eds) Field manual of techniques in invertebrate pathology. Kluwer Academic Publishers, Dordrecht, The Netherland, pp 255–282
10. Burges HD (1998) Formulation of mycoinsecticides. In: Burges HD, Ed HD (eds) Formulation of Microbial Biopesticides: Beneficial Microorganisms, Nematodes and Seed Treatments. Kluwer Academic Publisher, Dordrecht, pp 132–185
11. Butt TM, Jackson CW, Magan N (2001) Fungi as biocontrol agents: progress, problems and potential. CAB International, Wallingford, UK, pp 1–8

# Part V

## Mass Production and Formulations: Viruses

# Chapter 9

# Cell Culture for Production of Insecticidal Viruses

**Steven Reid, Leslie C.L. Chan, Leila Matindoost, Charlotte Pushparajan, and Gabriel Visnovsky**


## Abstract

While large-scale culture of insect cells will need to be conducted using bioreactors up to 10,000 l scale, many of the main challenges for cell culture-based production of insecticidal viruses can be studied using small-scale (20–500 ml) shaker/spinner flasks, either in free suspension or using microcarrier-based systems. These challenges still relate to the development of appropriate cell lines, stability of virus strains in culture, enhancing virus yields per cell, and the development of serum-free media and feeds for the desired production systems. Hence this chapter presents mainly the methods required to work with and analyze effectively insect cell systems using small-scale cultures. Outlined are procedures for quantifying cells and virus and for establishing frozen cells and virus stocks. The approach for maintaining cell cultures and the multiplicity of infection (MOI) and time of infection (TOI) parameters that should be considered for conducting infections are discussed.

The methods described relate, in particular, to the suspension culture of *Helicoverpa zea* and *Spodoptera frugiperda* cell lines to produce the baculoviruses *Helicoverpa armigera* nucleopolyhedrovirus, HearNPV, and *Anticarsia gemmatalis* multicapsid nucleopolyhedrovirus, AgMNPV, respectively, and the production of the nonoccluded *Oryctes* nudivirus, OrNV, using an adherent coleopteran cell line.


**Key words** Insect cell technology, Suspension culture, Adherent culture, Insecticidal viruses, Bioreactors

## 1 Introduction

The major viruses considered for use as insecticides are baculoviruses due to their stability in the outside environment as a result of their natural occlusion by polyhedra and their inability to infect vertebrates [1, 2]. While a number of examples of successful programs to produce baculovirus biopesticide products using infected larvae have been documented over the past 40 years [2–4], a major limitation to their wider use is the lack of a cost-effective in vitro production technology [1, 5].

To be cost competitive with larvae production for the major baculovirus biopesticide markets, the manufacturing of baculoviruses using cell culture will need to be conducted using cells that display a doubling time of 24 h or less when grown in suspension

Travis R. Glare and Maria E. Moran-Diez (eds.), *Microbial-Based Biopesticides: Methods and Protocols*, Methods in Molecular Biology, vol. 1477, DOI 10.1007/978-1-4939-6367-6_9, © Springer Science+Business Media New York 2016

culture using up to 10,000 l scale airlift or stirred tank bioreactors . Further, upon infection they will need to produce yields of at least 300–600 Occlusion Bodies per cell (OB/cell) at high cell densities (5–20 × $10^6$ cells/ml) [1]. This presents significant challenges but progress toward these production levels has been made with the three cell/baculovirus systems listed as follows.

The *H. zea* (HzAM1) cell line produced by Arthur McIntosh [6] has received a lot of attention for its ability to produce the *Helicoverpa armigera* nucleopolyhedrovirus, HearNPV, in vitro [7–9]. The Sf9 (*Spodoptera frugiperda*) cell line has been studied extensively in relation to its ability to be grown in suspension culture using serum-free media [10] and shows potential for the production of a *S. frugiperda* multicapsid nucleopolyhedrovirus, SfMNPV [11]. Finally a *S. frugiperda,* Sf21, and an *Anticarsia gemmatalis* cell line are available that show potential to produce an AgMNPV product [12, 13]. In this chapter, methods related to the suspension culture of *H. zea* and Sf21 cell lines to produce the HearNPV and AgMNPV viruses, respectively, will be provided.

However, insect cells that do not grow in suspension culture and that have slow growth rates (greater than 24 h doubling time) and nonoccluded viruses deserve attention as they may also meet future commercial opportunities. To cover the challenges of such cell lines and viruses, we include methods related to the production of a wild-type nonoccluded *Oryctes* nudivirus [14, 15], using a slow growing adherent coleopteran cell line, DSIR-HA-1179 [16].

The discovery and distribution of OrNV has been very successful in controlling the rhinoceros beetle pest of the palm oil industry [17]. Damage valued in millions of dollars annually is caused by the adult beetle which feeds on the growing shoots of palms often leading to death of the palm. However, control strategies based on the use of this virus suffer from difficulties in producing sufficient amounts of active virus, poor formulation of the virus to maintain its infectivity, and poor quality control in ensuring that only virulent strains of virus are released [18, 19]. The ability to produce OrNV in cell culture will overcome many current limitations in mounting an effective research program to more effectively use this virus as part of an Integrated Pest Management (IPM) strategy to control the beetle. While this virus will not need to be produced at as large scale as the HearNPV, SfMNPV, and AgMNPV viruses, it is likely that it will need to be produced using roller bottle or microcarrier-based systems and so these methods are outlined in this chapter using the DSIR-HA-1179 cell/OrNV system as a case study.

In this chapter there is insufficient space to address all the issues of insect virus production in cell culture. It is important to be aware that many of the issues of growing insect cells in bioreactors are similar to those faced for growing mammalian cells in culture. Hence the development of technology to maximize the production of monoclonal antibodies using CHO cells is relevant and of benefit for the large-scale growth of insect cells. The shear sensitivity and oxygen consumption properties of insect cells are similar to those of CHO and other mammalian cells grown in suspension culture and so reactor/impellor and air/oxygen sparging systems developed for producing mammalian cell-based products can be utilized for insect cell suspension cultures. The formulation of serum-free media used for insect cell cultures and feeds for fed batch processes are also similar to those used in mammalian cell processes. Indeed insect cell cultures can be easier to culture than mammalian cells in some respects in that they normally do not require pH control during batch or fed-batch runs as they tend not to produce the high lactate levels seen by many mammalian cells in culture.

Thus, many of the methods and protocols developed for animal cell biotechnology in general are applicable to insect cell culture processes and readers are encouraged to review other books in the "Methods in Molecular Biology" series relevant to Animal Cell Biotechnology [20]. Similarly books in this series focusing on Baculovirus and Insect Cell Expression Protocols [21] are directly relevant to the topic of producing insecticidal viruses in culture. The use of insect cells, particularly Sf9 and High-Five (*Trichoplusia ni*,T. ni), cell lines have been studied extensively in relation to their ability to produce various vaccine, therapeutic, and gene therapy products using baculovirus expression systems [22, 23], and this work has much to teach us about producing virus insecticides.

While large-scale culture of insect cells is not easy and requires the involvement of experienced animal cell technology personnel, virus yields from bioreactor insect cell cultures are comparable to yields obtained in small-scale insect cell shaker cultures [24]. The main challenges for cell culture-based production of insecticidal viruses relate to the development of appropriate cell lines, managing problems related to stability of the virus strains in culture, enhancing virus yields per cell through an understanding of how the host cell responds to the infecting virus, and the development of chemically defined media and feeds for the desired production systems [1, 2]. All of these issues can be studied using small-scale (20–100 ml) shaker/spinner flasks, particularly given the fact that many insect cells growing in suspension culture do not alter the pH of their growth medium and so do not require online pH control. Such simple suspension culture systems can operate to high cell densities (up to $10^7$ cells/ml) without showing signs of oxygen limitations [24, 25] and so there is no need to go to cumbersome bioreactor systems with pH and oxygen control to perform many of the required basic studies.

Hence this chapter presents mainly the methods required to work with and analyze effectively insect cell systems using small-scale suspension cultures. Outlined are procedures for quantifying cells and virus and for establishing frozen cell and virus stocks. The approach for maintaining stock cultures in good condition and the multiplicity of infection (MOI) and time of infection (TOI) parameters that should be considered for conducting infections are discussed. As baculoviruses are produced in two forms, budded virus (BV) and occlusion bodies (OB), methods for quantifying both these virus forms are shown. BV is the virus form produced when virions bud from an infected cell, in order to spread an infection via hemolymph from one caterpillar cell to another. The BV is also the form of the virus normally used to set up infections in cell culture. OBs are the form in which baculoviruses spread from 1 host insect to another and are stable in the outside environment due to the polyhedral protein coat surrounding the virus. This is the version of the virus that is used as an insecticide. OBs occlude what are referred to as ODV, occlusion-derived virus. ODV and BV are genetically identical but vary in the lipid and protein content of their membranes. For some viruses such as HearNPV, very poor BV titers are obtained from cell culture infections [26], and it is desirable to infect cells using ODVs and so the methodology to extract infectious ODV from OBs is also outlined.

As studies aimed at producing virus insecticides will eventually need to confirm that yields produced in shakers and spinner flasks can be reproduced in reactors, a basic method for growing insect cells in a stirred tank reactor (STR) and an airlift reactor (ALR) is included. Insect cells when infected by baculoviruses do become larger and develop a weakened cell membrane making them more susceptible to shear forces in a well-mixed bioreactor. Hence optimizing mixing conditions for large-scale cultures of infected insect cells may present some challenges beyond those met for mammalian cells. However, most current studies aimed at developing commercially viable processes involving insect cell cultures are based on systems biology approaches analyzing samples from small-scale suspension cultures [27–30].

## 2 Materials

### 2.1 Production of Insect Viruses in Suspension Cultures: Shake Flasks, Spinner Flasks, Stirred Tank, and Airlift Bioreactors

#### 2.1.1 Cell Line and Virus

1. The HzAM1 (Hzea) cell line used was derived from the pupal ovarian tissue of *Helicoverpa zea* (Lepidoptera: Noctuidae) (*see* ref. 6). This cell line was obtained from CSIRO, Division of Entomology, Canberra, Australia at passage number 242.
2. *Spodoptera frugiperda* cell line Sf-21 (ATCC CRL 1711), saUFL-AG-286 cell line (*see* ref. 12).
3. Wild-type *Helicoverpa armigera* nucleopolyhedrovirus, HearNPV, was obtained as caterpillar occlusion bodies, strain H25EA1, an Australian isolate, from CSIRO (Entomology Division, Canberra, Australia), which was used to infect the HzAM1 cell line.
4. Wild-type *Anticarsia gemmatalis* multiple nucleopolyhedrosis virus (AgMNPV) (*see* ref. 31).

*2.1.2 General Cell Culture Equipment and Consumables*

1. Incubator maintained at 27–28 °C.
2. Biological Safety Cabinet (Class II certified).
3. Pipet gun (e.g., Pipet-Aid®, Drummond Scientific).
4. 1, 2, 5, 10, 25, and 50 ml plastic disposable sterile pipettes.
5. 96 well plates, tissue culture treated, flat bottomed.

*2.1.3 Culture Medium and Supplements (See Note 1)*

1. TC-100 insect basal medium with l-glutamine and sodium bicarbonate, liquid, sterile filtered (Sigma).
2. IPL-41: basal medium (e.g., Life Technologies).
3. Fetal bovine serum (FBS), supplement (Life Technologies).
4. Chemically Defined Lipid Concentrate, supplement (e.g., Life Technologies).
5. Yeastolate Ultrafiltrate 50×, supplement (e.g., Life Technologies).
6. Antifoam A, supplement (Sigma).
7. Pluronic F-68, supplement (Sigma).
8. Sf-900™ II: liquid, complete serum-free media (Life Technologies).
9. Sf-900™ III: liquid, complete serum-free media (Life Technologies).
10. Sodium dodecyl sulfate (SDS).

*2.1.4 Bioreactors*

1. Orbital shaker platform.
2. Shaker Flasks: Erlenmeyer flask with screw cap (autoclavable glass and/or disposable (e.g., polycarbonate)), 125 and 250 ml.
3. Spinner Flasks: 500 ml glass spinner flask with pendular magnetic agitator, mounted on magnetic stirrer base (Techne, UK).
4. Stirred Tank Reactor: 5 l Biostat® A glass fermentor (B. Braun Biotech, Melsungen, Germany).
5. Airlift Reactor: 1.0 l glass GAV-3 concentric tube airlift reactor.

*2.1.5 Cell Counts/ Sampling*

1. Manual cell counts: phase-contrast microscope (e.g., BX43 upright microscope, Olympus), improved-Neubauer hemocytometer, and 2-key manual cell counter (i.e., click counter).
2. Automated cell counts (e.g., Multisizer™ 4 Coulter Counter or ViCell® Cell Viability Analyzer, Beckman Coulter).
3. Diluent for cell counts (same medium as that used to propagate cells).
4. Trypan Blue 0.4% (w/v) solution.

*2.1.6 Cell Cryopreservation Equipment*

1. Cryogenic dewar (e.g., LD50, 50 l capacity, liquid nitrogen storage, Taylor-Wharton).
2. Ultralow temperature freezer (e.g., VIP™ MDF-U55V, −86 °C, Panasonic).

3. Freezing container (e.g., Mr Frosty™ Freezing Container, Thermo Fisher Scientific).
4. Cryogenic vials (e.g., 1 ml Nunc Cryotubes™, Thermo Fisher Scientific).
5. Cryogenic vial protection (e.g., Nunc Cryoflex™ Tube Wrap, Thermo Fisher Scientific).
6. Spirit lamp and scissor (hemostatic) clamps.
7. Isopropyl alcohol (≥99.7%).
8. Dimethyl sulfoxide (DMSO) (sterile-filtered, ≥99.7%, e.g., Hybri-Max™, Sigma-Aldrich) (*see* **Note 2**).

### 2.2 Production of Cells and Virus in Adherent Cell Cultures: T-Flasks, Roller Bottles, Microcarriers

#### 2.2.1 Cell Line and Virus

1. The DSIR-HA-1179 coleopteran cell line originally was derived from embryonic tissues of the black beetle *Heteronychus arator* [16]. The cell line can be obtained from CAPE cell culture lab, University of Canterbury, New Zealand, where it is routinely maintained at 27 °C as adherent cultures (*see* ref. 32).
2. Wild-type *Oryctes* nudivirus (strain X2B) was produced in vitro in infected DSIR-HA-1179 cell cultures, harvested and stored at 4 °C. The X2B strain was originally isolated from a field population of infected coconut rhinoceros beetles on Bugsuk Island, Palawan, Philippines in 1983 (*see* refs. 14, 15). OrNV stock may also be obtained from CAPE cell culture lab (*see* ref. 32).

#### 2.2.2 Culture Medium, Solutions, and Microcarriers

1. TC-100 insect medium with l-glutamine and sodium bicarbonate, liquid, sterile filtered (Sigma) (*see* **Note 3**).
2. Fetal bovine serum (FBS) (Life Technologies).
3. TrypLE™ Express (Life Technologies).
4. Dulbecco's phosphate buffered saline (D-PBS) free of calcium and magnesium (Sigma).
5. Cytodex-1 microcarriers (Sigma).
6. Trypan blue dye (0.4%) (Sigma).
7. Crystal violet lysis buffer: 0.9 g citric acid powder (Sigma) is dissolved in 100 ml distilled $H_2O$ and mixed well. 0.01 g of crystal violet (Sigma) is added to the mixture and homogenously dispersed.

#### 2.2.3 Bioreactors

1. T-Flasks: 25 and 75 $cm^2$ T-flasks with plug seal caps (Corning).
2. Roller Bottle System: (a) CELLROLL roller bottle system, consisting of two roller racks mounted with a drive unit (0.1–2 rpm) and connected to a control unit and drive supply (Integra Biosciences), (b) 490 $cm^2$ roller bottles with plug seal caps (Corning).
3. Spinner Flasks: 125 ml glass spinner flasks with pendular magnetic agitator, mounted on magnetic stirrer base (Techne).

# 3 Methods

## 3.1 Production of Insect Viruses in Suspension Cultures

As indicated in the Introduction, this section will concentrate on methods related to the production of the baculoviruses HearNPV and AgMNPV in suspension cultures (*see* **Note 4**).

### 3.1.1 Cell Density and Cell Viability Enumeration

Cell densities are usually determined via a counting chamber (Improved Neubauer hemocytometer) and a phase-contrast microscope, as follows:

1. Estimate the dilution required so that approximately 100 cells are counted on each side of the hemocytometer (each side being a counting grid with nine large squares, each being 1 mm in length and 0.1 mm in depth, with a cover slip).
2. Perform 1:1 serial dilutions of the culture sample with medium, but use 0.1 % (w/v) Trypan Blue (prepared in medium) as the diluent for the final dilution.
3. Load each side of the hemocytometer (with cover slip) with 10 μl of the diluted sample, and count the number of cells on each side with a click counter under the microscope (count unstained and stained cells separately).
4. Tally up the number of viable (unstained) and total cells (unstained + stained) from both sides of the hemocytometer (18 squares).
5. Calculate the viable and total cell density using the following formula: cell density (cells/ml) = (no. of cells in 18 squares × dilution factor)/0.0018 ml.
6. If the cell density is estimated from 3× hemocytometer counts (~600 cells counted from 54 large squares), then a relative error of approximately 15 % is obtained (*see* ref. 33).

Alternatively, viable and total cell densities may be determined using an automated imaging counting system, e.g., Cedex or ViCell (this can also determine the total cell density), and total cell densities may be determined using a Coulter Counter (e.g., Multisizer 4), according to the manufacturer's instructions. When using an imaging counting system, the analytical settings optimized for uninfected cells may not work well for infected cells (e.g., underestimated cell viabilities), in which case such settings should be reoptimized.

### 3.1.2 Cell Cryopreservation: Freezing/Thawing

Insect cells adapted to Serum-Free Media (SFM) can be frozen in liquid nitrogen for long-term storage using the following procedure, which works well for Sf-9, HzAM1, and Tn-5 cells in our laboratory. The general principle of animal cell cryopreservation is to freeze slowly and thaw quickly.

Freezing:

1. Prepare a 100 ml suspension culture, seeded at $5 \times 10^5$ cells/ml in a 250 ml shaker flask.
2. Prepare 15 % (v/v) DMSO in fresh culture medium in a 10 ml centrifuge tube and store at 4 °C.
3. When the cells are at mid-exponential growth phase (e.g., $2 \times 10^6$ cells/ml, >95 % viability, 24 h doubling time), aliquot 10 ml culture into a 10 ml centrifuge tube, centrifuge at $100 \times g$ for 5 min, transfer the supernatant into a new 10 ml tube and chill on ice (conditioned medium). Also chill the tube of 15 % (v/v) DMSO in fresh medium on ice. Perform **step 3** at ~1 h before **step 5**.
4. Label 1 ml cryogenic vials with the designated freeze numbers.
5. Aliquot the remaining culture (90 ml) into 2× 50 ml centrifuge tubes, centrifuge at $100 \times g$ for 5 min, and discard the supernatant.
6. Resuspend the cell pellet in the required volume (9 ml in 10 ml tube) to obtain a cell density of $2 \times 10^7$ cells/ml (tenfold concentration), e.g., 4.5 ml ice-cold 15% (v/v) DMSO in fresh medium and approximately 4.5 ml ice-cold conditioned medium (use the pipette's graduation to estimate the final volume of 9 ml).
7. Aliquot 1 ml cell concentrate into each cryogenic vial, ensuring that the 10 ml tube is well mixed between each dispensing step, and chill on ice.
8. Insert each vial into a short length of Cryoflex™ Tube Wrap, and seal each end of the Cryoflex with the aid of a spirit lamp and scissor clamps.
9. Install the Cryoflexed vials in a 'Mr Frosty' freezing container filled with isopropyl alcohol, and place in an ultralow temperature freezer (−80 °C). The 'Mr Frosty' container ensures a slowed-down freezing rate of −1 °C/min.
10. On the next day, remove the vials from the 'Mr Frosty', install them on appropriate freezing canes, and store them under liquid nitrogen in a cryogenic dewar.

Thawing:

1. Remove a vial from the cryogenic dewar, and place it on dry ice for transport.
2. Rapidly thaw the frozen cells by placing the vial in a 28 °C water bath. Swirl the contents gently to speed up thawing.
3. In a BSC resuspend the thawed cells and transfer into a 10 ml centrifuge tube and add 5 ml fresh SFM.

4. Centrifuge the tube at 100 × *g* for 5 min, discard the supernatant, and resuspend the cell pellet in fresh SFM to a final volume of 25 ml (~$8 \times 10^5$ cells/ml) in a 125 ml shaker flask (passage 1 after thaw).
5. Incubate the cells and monitor cell growth, viability, and sterility daily. Maintain using the cell passaging routine as described in Subheading 3.1.4 (*see* **Note 5**).

#### 3.1.3 Aseptic Technique in the Biological Safety Cabinet (BSC)

1. Aseptic technique is critical for insect cell cultures as the risk of microbial contamination is high due to the richness of the growth media and the slow growth rate of insect cells in comparison to that of most microbes (24 h vs. ≤1 h doubling times). Insect cell culture experiments, uninfected or baculovirus infected, can take up to 2–3 weeks to complete depending on scale and mode (batch or fed batch).
2. The sterility of insect cell cultures is best managed inside a Class II BSC and by using presterilized single-use plastic accessories (e.g., culture flasks, serological pipettes, filters, bottles, and centrifuge tubes) as much as possible. If reusable items are employed, then these should be well cleaned, depyrogenized, and autoclave sterilized.
3. Before working in a BSC, the cabinet bench and any items placed on it should be surface sanitized to reduce the microbial load [e.g., by wiping with paper towels soaked in 70% (v/v) ethanol].
4. The laminar flow should be started at least 30 min prior to work.
5. When performing liquid handling procedures in a BSC, a high degree of attention to detail is required to maintain asepsis: (a) HEPA-filtered sterile air flows from the top to the bottom of the cabinet; thus, a sterile item (liquid or solid) remains sterile if care is taken not to pass a nonsterile item above it and physical contact between the sterile item and any nonsterile entity is avoided (solids, liquids, gases, or aerosols). (b) Good aseptic technique includes having a noncluttered work area, leaving sufficient empty space between sterile objects, and a spatial memory of where the sterile and nonsterile items are situated.
6. The BSC should be decontaminated and NATA tested annually to maintain optimal performance.

#### 3.1.4 Culture of HearNPV in Shaker Flasks

Insect cell suspension cultures are initiated from cryopreserved stock cells. Once thawed, the cells are serially passaged regularly (e.g., twice weekly) in fresh Serum-Free Media (SFM) as Erlenmeyer shaker flask batch cultures. The following cell passaging procedure is suitable for Hzea cells grown in an optimized SFM such as

Sf-900 III. Small-scale 125 ml shaker flasks (20–50 ml working volume) are used to save on medium costs. If larger volumes are required, then 250 ml shaker flasks (50–100 ml working volume) can be used.

1. Hzea stock cells are ready for passaging when they have reached the mid-exponential growth phase (medium dependent). Hzea cultures grown in Sf-900 III reach a Peak Cell Density (PCD) of $8 \times 10^6$ cells/ml or better in our hands with a mid-exponential density of around $4 \times 10^6$ cells/ml with a cell viability of >95 % (*see* **Note 6**).
2. On the passaging day set up a new 25 ml culture (in a 125 ml flask) at a seeding density of $4–5 \times 10^5$ cells/ml, which represents an approximate tenfold dilution of the stock cells with fresh Sf900III (e.g., 2.5 ml cells + 22.5 ml SFM).
3. Install the flask on an orbital shaker platform (120 rpm) in a refrigerated incubator (27–28 °C). Ensure that the screw cap is loosened (e.g., quarter turn anticlockwise) to allow for gas exchange.
4. Allow the new stock cells to reach mid-exponential growth phase and then repeat the cell passaging procedure.
5. For a twice-weekly passaging routine, the following schedule works well for us (assuming a 24 h cell doubling time): *Monday*: Set up stock cells at $2.5–3 \times 10^5$ cells/ml, *Friday*: Stock cells grown to $4–5 \times 10^6$ cells/ml (Passage N), set up new stock cells at $4 \times 10^5$ cells/ml, *Monday*: Stock cells grown to $3–4 \times 10^6$ cells/ml (Passage N + 1), set up new stock cells again at $2.5–3 \times 10^5$ cells/ml, and so forth.
6. The procedure used to setup experimental batch cultures is the same as that used for serial passaging of stock cells. The same agitation speed can be applied for both 125 ml and 250 ml shaker flasks (120 rpm).
7. If the batch cultures are to be infected with a baculovirus, then the settings of certain key infection parameters have to be optimized, including the infection cell density (ICD), the multiplicity of infection (MOI), and the Peak Cell Density (PCD).
8. For the production of HearNPV occlusion bodies (OBs) using Hzea cells in Sf900III, typical settings for ICD and MOI are $2–4 \times 10^6$ cells/ml and 5–10 PFU/cell, respectively. Maximum volumetric yields are obtained by infected cultures that reach PCDs of $3–4 \times 10^6$ cells/ml (*see* **Note 7**).

#### 3.1.5 ODV Extraction to Initiate HearNPV Infections in Suspension Culture

Baculoviruses are genetically unstable when passaged in culture due to the generation of defective interfering particles, DIPs (*see* ref. 34), and the rapid accumulation of few polyhedral, FP, mutants (*see* ref. 35). Normally to infect cells in culture with new wild-type baculovirus isolates, infected caterpillars are harvested in nature, and extracted hemolymph is diluted with media, filter sterilized and the BV so obtained are used to infect cells in culture. However, this approach makes it difficult to establish master and working stocks of virus in sufficient volumes to support a large-scale manufacturing process, particularly if the BV yields in culture are low as is the case for HearNPV produced in culture (*see* ref. 26).

One solution to this problem is to maintain master and working stocks of baculoviruses as caterpillar generated OBs and extract the ODVs from the OBs to set up a BV stock in culture. However, ODV have been reported as only infecting cells in static cultures very poorly (*see* ref. 36). For HearNPV, we have found that ODVs extracted from OB infect Hzea cells in suspension culture efficiently and a commercially viable process is feasible that relies on the use of master and working stocks of OBs stored at 4 °C or frozen at −80 °C (*see* ref. 1). A method to extract ODVs for infecting cells in suspension culture is outlined as follows.

1. Take 500 μl of an OB stock generated in caterpillars and stored at a concentration of $10^{10}$ OB/ml in water or media at 4 °C or frozen at −80 °C.
2. Add 40 μl of an alkali solution (0.5 M $Na_2CO_3$ and 1.0 M NaCl) to the OBs in an eppendorf tube, vortex and incubate for 30 min at 28 °C.
3. Mix the digested OBs with 10 ml of Sf900III medium to neutralize the extract.
4. In a sterile cabinet use a 10 ml sterile syringe to suck up the ODV solution and filter it through a sterile 0.22 μm filter (Durapore® PVDF, Sartorius, Australia) into a sterile 10 ml tube (*see* **Note 8**).
5. Add this ODV extract (~9.5 ml) to a 90 ml culture (250 ml shaker flask) such that the final cell density at the time of infection is $5 \times 10^5$ cells/ml.
6. Incubate the infected culture at 28 °C and 120 rpm for 4 days to produce a P1 virus stock.
7. Use the P1 virus to immediately generate a P2, BV stock, by adding the P2 whole culture, 30 % (v/v), to cells such that the final cell density at the time of infection is $1 \times 10^6$ cells/ml.
8. At 3 days postinfection harvest the P2, BV supernatant by centrifugation, $1000 \times g$ for 10 min at room temperature and store for 2–4 weeks at 4 °C or at −80 °C if to be stored for longer periods.
9. The P2, BV virus stock produced in this manner will typically have a titer of $2–5 \times 10^7$ PFU/ml (*see* **Note 9**).

10. All experiments aimed at optimizing yields ideally will be done using P2 virus stocks to initiate infections as excessive passaging of virus in culture can significantly affect the OB/cell yield (*see* ref. 35). A commercially viable process is feasible taking this approach whereby the virus will only be in cell culture for a total of three passages, counting the final fed batch production run (*see* ref. 1).

*3.1.6 OB Count*

1. Extract OB for counting by mixing an equal volume of 1% SDS to 0.2 ml of an infected cell suspension (*see* **Note 10**).
2. Incubate at 28 °C for 30 min to allow dissolution of the cell membrane and release of the polyhedra.
3. Dilute the OB extract serially with pure water to allow a count of 100–150 OB per small square of a hemocytometer.
4. A 0.1-mm-deep Improved Neubauer hemocytometer (Weber Scientific International Ltd., England) which consists of two chambers each divided into nine 1 mm squares is to be used for counts. The central square is further divided into 25 smaller squares which is the square used for this particular count. The volume of each 1 mm square, including the central square is equal to $10^{-4}$ ml.
5. Clean the counting chamber and cover-slip with 75% ethanol, dry with a paper wipe, and fix cover-slip in position.
6. A total of 5 small squares (out of the 25 smaller squares within the central square) were counted on each side of the hemocytometer (count OBs in the four corner plus the central small square). Each sample was counted using three hemocytometers, both sides, using an optical microscope (Olympus, Japan) at 400× magnification (including 10× eyepieces and 40× objective lens). Diluted samples were counted 10 min after loading to allow the polyhedra to settle on the base of the hemocytometer.
7. Each count consists of a tally of the number of occlusion bodies completely contained within a small square plus the number touching the left-hand and upper sides whereas occlusion bodies touching the bottom and right-hand sides are not counted.
8. The polyhedra concentration was then determined as follows (per side of each hemocytometer count:

$$\text{OB} / \text{ml}\left(\text{one side}\right) = \text{number counted in 5 small squares} \times 5 \times \text{dilution factor} \times 10^{4}.$$

9. The volumetric OB yield of each sample was the average of the OB/ml count from each side (six sides in total) of the hemocytometer.
10. A count conducted is this manner counts a minimum of 600 OBs and results in an acceptable random error of 15% (*see* ref. 33).

11. The cell-specific yield (polyhedra per cell) is obtained by dividing the volumetric yield (OB per ml) by the total peak cell density (*see* **Note 11**).

*3.1.7 Culture of AgMNPV in Spinner Flask Bioreactors*

1. Sterilize a 500 ml spinner flask and place it into the laminar flow cabinet.
2. Prepare the culture medium by adding fetal bovine serum to the TC-100 culture medium at a concentration of 10%.
3. Prewarm the culture medium in the incubator at 27 °C and aseptically transfer 50 ml of culture medium into the spinner flask.
4. Inoculate the spinner flask with an appropriate volume of IPLB-Sf-21 cells inoculum to give an initial cell density of $2 \times 10^5$ viable cells/ml.
5. Transfer the spinner flask to the magnetic stirrer base held within the incubator and adjust the speed of the magnetic stirrer to 60 rpm.
6. Withdraw 1 ml samples of the cell suspension in a BSC with a 1 ml pipette at regular intervals in order to assess viable cell density and culture viability.
7. Infect the culture, for example, during the early exponential growth phase by adding an appropriate volume of virus inoculum stock to obtain the desired MOI. Return the spinner flask to the incubator until 150 h postinfection when peak virus yields are obtained. Cell density at the moment of the infection, multiplicity of infection (MOI), and harvesting time could change according to experimental objectives (*see* ref. 31).
8. To assess budded virus titer, aseptically transfer 1 ml of an infected sample from the reactor into an Eppendorf tube, and centrifuge it at 4000 × *g* for 10 min. The clear supernatant contains the budded virus progeny, which can either be stored at 4 °C or directly quantified according to the method in Subheading 3.2.6.
9. Polyhedra yields can be determined as described in Subheading 3.1.6.

*3.1.8 Culture of HearNPV in Stirred Tank Bioreactors*

1. The 5 l bioreactor was prepared for cell culture by cleaning and depyrogenization using 10 g/L Terg-a-zyme® detergent (Alconox, White Plains, NY) and 0.1 M NaOH, respectively (overnight soaking), followed by rinsing with deionized water, assembly and autoclave sterilization for 45 min (121 °C, 100 kPa). Hydrophobic filters (0.2 μm Millex FG-50, Millipore) were installed for gas addition or venting.
2. The bioreactor was operated at an agitation speed of 160–220 rpm, to account for changes in liquid volume during fed-batch processes, while maintaining a constant ungassed power/volume (P/V) ratio of 16.3 W $m^{-3}$. The power consumption

was estimated by using a well-established correlation with agitation speed (*see* ref. 37) and a power number of 1.7 for pitched blade 'elephant ear' impellers (*see* ref. 38).

3. Bioreactor cultures were maintained at a temperature set point of 28 °C via the micro DCU-400 control system (B. Braun Biotech).
4. The dissolved oxygen tension (DOT) was controlled at 50 % of air saturation using the Wheaton Control Tower® (Wheaton Science Products, Millville, NJ) with pure $O_2$ sparging. In addition, a low flowrate of air was introduced into the head-space gas inlet.
5. The liquid addition/withdrawal device was either a glass bottle or a fernbach flask, adapted with a vented screw-cap lid (fitted with a 0.2 μm hydrophobic filter) and a glass spigot at the base (fitted with a length of silicone tubing and ending with a polypropylene Y-piece connector). Similar tube /Y-piece assemblies were also fitted to the bioreactor's addition/withdrawal ports. During autoclaving, the tubes were sealed using gate clamps just upstream from the Y-piece. Two lengths of tubing were connected aseptically by fitting one arm of each Y-piece together with tubing and connecting the other arm to either a live steam source or a steam trap. The Y-piece assembly was then steamed for 30 min (180 kPa supply pressure), then the steam and condensate lines were clamped, and the connection was cooled prior to use.

#### *3.1.9 Culture of AgMNPV in Airlift Bioreactors*

1. Sterilize a 1.0 L glass concentric airlift reactor and place it into the laminar flow cabinet.
2. Fill it partially (~50 % of its working volume) with prewarmed at 27 °C TC-100 culture medium supplemented with 10 % fetal bovine serum, 200 ppm silicone antifoam and 0.20 % w/v of Pluronic F-68.
3. Inoculate the reactor with an appropriate volume of IPLB-Sf-21 cells inoculum grown in a spinner flask to give an initial cell density of $2 \times 10^5$ viable cells/ml.
4. Complete reactor working volume by adding culture medium as described in Subheading 3.1.9 **step 2** making sure its level goes over the reactor's riser (internal draft tube).
5. Check outlet of reactor is open and protected by a 0.2 μm air filter.
6. Connect the reactor to the air supply source and sparge air filtered with 0.2 μm filter into the airlift reactor at a superficial gas velocity, $J_G$, ranging from 0.09 to 0.1 cm/min.
7. Withdraw 1 ml samples of the cell suspension. The system used for this task varies according to the airlift reactor design. Usually, a sample is collected from a secondary port which has a vial

attached by exerting a slightly positive pressure inside the reactor (or negative pressure from the vial). Once the sample is taken, the vial is replaced with a new clean one for the next sample.

8. Infect the culture, for example, during the early exponential growth phase by adding an appropriate volume of virus inoculum stock to obtain the desired MOI. Cell density at the moment of the infection, multiplicity of infection (MOI), and harvesting time could change according to experimental design and objectives (*see* refs. 12, 39).
9. To assess budded virus titer, aseptically transfer 1 ml of infected sample from the reactor into an Eppendorf tube, and centrifuge it at 4000 × *g* for 10 min. The clear supernatant contains the budded virus progeny, which can either be stored at 4 °C or directly quantified according to the method in Subheading 3.2.6.
10. Polyhedra yields can be determined as described in Subheading 3.1.6.

### 3.2 Production of Insect Viruses Using Adherent Cultures: DSIR-HA-1179-OrNV System as an Example

#### 3.2.1 Preparation of Culture Medium

1. Ensure that the culture medium is free of adventitious agents. It must always be stored in sterile, capped bottles. Glassware must be soaked overnight in a solution of 1 % Virkon, rinsed well with MilliQ water and autoclaved (121 °C for 20 min) prior to use.
2. The DSIR-HA-1179 cell line requires 10 % FBS for optimal growth. Therefore, prepare appropriate volumes of TC-100 culture medium supplemented with 10 % FBS (*see* **Note 12**).
3. Prepare culture medium ahead of use. Incubate it at 27 °C for 24 h to check for microbial contamination prior to use.

#### 3.2.2 Preparation of Cell Inoculum

1. Prepare cell inoculum to seed bioreactors from T-flask cultures in which the cell monolayer is 80–90 % confluent.
2. Pipette out spent culture medium from the T-flask and discard.
3. Add 2 ml of D-PBS free of calcium and magnesium per 25 $cm^2$ of flask surface area. Rock the flask gently to evenly coat the surface for 2 min. Pipette out the spent D-PBS and discard.
4. Add 1 ml of TrypLE™ Express (prewarmed to 27 °C) per 25 $cm^2$ of flask surface area. Rock the flask gently to evenly coat the surface. Transfer flask to the incubator and incubate for 30 min at 27 °C.
5. Observe culture under microscope to confirm cells have detached. Add an appropriate volume of prewarmed culture medium to the flask and pipette gently to break up any cell clumps and to create a homogenous single cell suspension (*see* **Note 13**).
6. Transfer 1 ml of the cell suspension to a 1.5 ml microcentrifuge tube to make a cell count. A sample of the cell suspension is

stained with trypan blue and loaded on the hemocytometer. Both total and viable cells are counted in duplicate in order to estimate viable cell density and culture viability (*see* refs. 40, 41).

7. Based on viable cell count, use the appropriate volume of culture to inoculate the bioreactor.

#### 3.2.3 OrNV Production in T-Flasks

1. Inoculate a 25 $cm^2$ T-flask at an initial DSIR-HA-1179 cell density of $2 \times 10^5$ viable cells/ml in a culture volume of 5 ml (*see* **Note 14**).
2. Incubate the culture at 27 °C until early exponential growth phase (cell density of ~$5 \times 10^5$ viable cells/ml).
3. Infect the culture by adding to it the appropriate volume of OrNV stock to achieve the desired MOI (*see* **Note 15**).
4. Incubate the infected culture at 27 °C until day 6 postinfection (*see* **Note 16**).
5. In order to harvest virus, transfer the entire contents of the infected culture into a 15 ml centrifuge tube and centrifuge at $4000 \times g$ for 10 min (*see* ref. 32).
6. Aseptically transfer the supernatant (containing virus) to a fresh, sterile centrifuge tube and store at 4 °C.

#### 3.2.4 OrNV Production in the Roller Bottle System

1. Add 25 ml of fresh culture medium to the roller bottle and roll at 0.1 rpm for 24 h in the incubator at 27 °C to precondition the surface of the roller bottle (*see* **Note 17**).
2. Inoculate the roller bottle with an appropriate volume of cell inoculum to obtain a cell density ~$4 \times 10^4$ cells/$cm^2$. Incubate roller bottle at 0.1 rpm for a further 24 h. The inoculation is done in a reduced culture volume (25 ml) in order to facilitate better adhesion of the cells to the roller bottle surface and to form an even cell monolayer (*see* ref. 42).
3. Adjust the final culture volume to 60 ml by adding fresh culture medium to the roller bottle and continue incubation under the same conditions.
4. Cell growth is assessed by harvesting the full content of the roller bottle. It involves the dissociation of the cell monolayer with TrypLE™ Express enzyme using a slightly modified method to that described in Subheading 3.2.2, as follows: Pipette the spent culture medium out of the roller bottle and add 20 ml of D-PBS free of calcium and magnesium to the cell monolayer. Return roller bottle to incubator and roll at 0.1 rpm for 10 min. At the end of this period, remove the spent D-PBS and add 20 ml of TrypLE™ Express to the roller bottle. Return the bottle to the incubator and roll at 0.1 rpm for 30 min until cells have detached from the monolayer. Add an appropriate volume of prewarmed culture medium supplemented with 10%

FBS, and gently aspirate the cell suspension with a 25 ml pipette to break up any cell aggregates. Aseptically transfer a 1 ml sample of the cell suspension into a microcentrifuge tube and estimate cell density and culture viability.

5. Infect the roller bottle culture with OrNV at, for example, the early exponential growth phase (approximately $5 \times 10^5$ viable cells/ml) by adding an appropriate volume of virus inoculum stock to obtain the desired MOI. Return the roller bottle to the incubator and roll at 0.1 rpm until day 8 postinfection.
6. Harvest virus from the infected roller bottle culture on day 8 postinfection as peak virus yields are reached at this time point (*see* ref. 43). Gently transfer the entire content (60 ml) of the infected roller bottle culture and aliquot into equal volumes of 15 ml in four 15 ml centrifuge tubes. Centrifuge the tubes at $4000 \times g$ for 10 min. Aseptically transfer out the supernatant containing the virus into new sterile tubes and store at 4 °C.

*3.2.5 OrNV Production in Microcarriers*

1. Prepare a Cytodex-1 microcarrier stock solution of 10 g/l. Weigh 1 g of dry Cytodex-1 microcarriers and add it to 100 ml of D-PBS free of calcium and magnesium in a Schott-duran bottle. Gently swirl the mixture for 5 min to evenly disperse the microcarriers. Incubate the mixture at 27 °C for 5 h to hydrate the microcarriers.
2. Decant the supernatant and wash the microcarriers twice, with two changes of 50 ml of fresh D-PBS free of calcium and magnesium.
3. Autoclave the microcarriers in 50 ml of fresh D-PBS at 121 °C and 15 psi for 20 min. Store at 4 °C until time of use (*see* refs. 44, 45).
4. Transfer to the laminar cabinet and decant the D-PBS solution from the bottle. Wash the microcarriers with two exchanges of 50 ml serum-free TC-100 culture medium, prewarmed to 27 °C. Finally, resuspend microcarriers in 10 % FBS-supplemented TC-100 culture medium.
5. For the DSIR-HA-1179 cell line, a procedure of initially inoculating cells on microcarriers on a flat surface of a 75 cm² T-flask in a reduced culture volume of 20 ml is used, followed by a 12 h period under static conditions, which has been found to improve cell attachment. Following this, the culture is transferred to a spinner flask with a final culture volume of 60 ml.
6. Hence: Transfer an appropriate volume of microcarriers to give a concentration of 1 g/l for a final culture volume of 60 ml into a 75 cm² T-flask. Adjust the volume of the flask to 20 ml by adding fresh culture medium. Inoculate accordingly to obtain 30 cells per bead (microcarrier). Incubate the culture at 27 °C for 12 h under static conditions (*see* **Note 18**).

7. At the end of the 12 h attachment period, microscopically observe cultures to check that all cells have attached to microcarrier beads. Aseptically transfer the 20 ml culture out of the T-flask and into a sterile 125 ml spinner flask, and adjust the final culture volume by adding 40 ml of prewarmed culture medium.
8. Place the spinner flask onto the stirrer base within a 27 °C incubator. Adjust the magnetic stirrer speed to 40 rpm.
9. Total cell density in microcarrier cultures is evaluated at periodic intervals over batch growth by nuclei counting (*see* refs. 46, 47). Aseptically remove 1 ml samples of microcarrier culture into 1.5 ml microcentrifuge tubes and centrifuge the tubes at 12,000 × *g* for 5 min to separate cell-bound microcarriers from the culture supernatant. Remove the culture supernatant and add 1 ml of crystal violet lysis buffer to the cell-microcarrier pellet. Vortex the mixture for 1 min, and then incubate it overnight at 27 °C. Citric acid does not affect the microcarriers, and only lyses the cells' plasma and nuclear membranes to release their nuclei, which are stained by crystal violet. The individually stained nuclei are counted in duplicate using a Neubauer hemocytometer in order to assess total cell density (*see* ref. 40).
10. Infect the microcarrier culture with OrNV in the early exponential growth phase by adding an appropriate volume of virus inoculum stock to obtain the desired MOI. Return the spinner flask to the incubator until day 4 postinfection when peak virus yields are obtained.
11. To harvest the virus, aseptically transfer the entire 60 ml infected culture as 15 ml aliquots in four centrifuge tubes and centrifuge at 4000 × *g* for 10 min.
12. The cells and microcarriers will have pelleted at the bottom of each tube leaving a clear supernatant fluid containing the virus. Aseptically transfer the viral supernatant out into new sterile tubes and store at 4 °C.

#### *3.2.6 Quantification of Infectious Virus Titer by Endpoint Dilution (See Note 19)*

1. Prepare a single cell suspension of DSIR-HA-1179 cells from a parent T-flask culture using TrypLE™ Express treatment, as described in Subheading 3.2.2.
2. Dilute the culture with fresh culture medium to a concentration of $2.5 \times 10^5$ viable cells/ml.
3. Inoculate the wells in each of 5 columns of a 96-well plate with 50 μl of the cell suspension per well (*see* **Note 20**).
4. Set up a tenfold dilution series ($10^{-1}$–$10^{-9}$) of the virus sample in culture medium in microcentrifuge tubes.
5. From each dilution of viral supernatant, add 50 μl of the respective supernatant to each of five replicate wells in the 96-well plate.

6. Place the plates in a humidified, disinfected plastic container and incubate at 27 °C for 11–14 days until the cytopathic effect is well developed (*see* **Note 16**), and the plates can be reliably scored for infection. The incubation length to assess cytopathic effect changes with the insect cell line and virus used.
7. The $TCID_{50}$ value is calculated according to the method of Reed and Muench (*see* ref. 48).

## 4 Notes

1. The culture media described in Subheading 2.1.3 were used to culture the cell lines described in the section; however, the methodology applied is of general use for any insect cell line to be cultivated under the same conditions.
2. Dimethyl sulfoxide (DMSO) may be sourced from a different supplier.
3. Alternatively TC-100 dry powder medium (Sigma or Invitrogen) may be purchased and prepared for use according to the manufacturer's instructions.
4. While a case was made in the introduction that the current challenges for producing baculovirus insecticides in culture (yield/media/feed improvements) can be addressed at small scale—the small-scale work referred to is 20–100 ml suspension shaker/spinner cultures. Research in static cultures is not as relevant to large-scale suspension processes. Static cultures are not mixed and oxygen limitation becomes an issue once cells get above $1–2 \times 10^6$ cells/ml. In addition, adherent cells most likely have a different protein expression profile for many proteins compared to that for cells in suspension and it is very difficult to accurately quantify cells in static cultures leading to inaccurate quantification of cell-specific virus yields.
5. Cryopreserved cells should have recovered (i.e., normal cell growth rate and viability) by 2–3 passages after thaw. If not, then discard the cells and thaw out another vial from the cell bank. If there is a low success rate of recovering cells from a particular cryopreserved batch, then the freezing procedure may not have been carried out properly. In this case, repeat the cryopreservation procedure with a new batch of cells.
6. Insect cells will obviously exhibit different growth characteristics depending on the medium used. One of the most important initial tasks in insect cell culture is to establish a cell growth curve (cell density and viability over time) for a particular medium, from which the cell doubling time, mid-exponential growth phase, and PCD can be determined. This information is used to set the cell density and temporal parameters for passaging events.

7. Peak cell-specific yields of HearNPV of 400–500 OB/cell are obtained only for infections at 0.5–1 × $10^6$ cells/ml. The cause of the drop off in peak cell specific yields seen for infections conducted at higher cell densities is due to the so-called cell density effect (*see* ref. 49). Cell-specific and volumetric yields at higher cell densities can be improved by the use of fed-batch processes (*see* refs. 24, 25). It is also possible to develop low cost media that give similar yields to those obtained with commercial media such as Sf900III (*see* refs. 50, 51). Low cost media for insect cells are typically based on the basal medium formulation of IPL41, plus a lipid emulsion additive containing cholesterol and a yeast extract. Other hydrolysates are often required if the addition of expensive purified amino acids is to be avoided (*see* ref. 50). Feeds for insect cells are based on concentrates of the ingredients used in the media. Quality assurance is a significant problem in producing low cost media if they are to perform in a reproducible manner, due to the challenges of producing good quality lipid emulsions and variability in the quality of the hydrolysates and the yeast extracts used.
8. Typically it requires two disposable filters to sterilize the 10 ml ODV extract.
9. The P2, BV titer can be determined using a plaque assay, an endpoint titration assay (*see* Subheading 3.2.6), or by a relatively simple suspension culture-based assay (*see* ref. 52). BV can be stored at 4 °C for short periods (up to 1–3 months) but for longer term storage they should be frozen quickly and stored at −80°°C or in liquid nitrogen (*see* ref. 53). BV should not be stored at −20 °C as at this temperature the BV freeze too slowly and are damaged (*see* ref. 53). Long-term storage at 4°C leads to loss of BV activity due to clumping of the virus according to an excellent study by Jorio et al. (*see* ref. 53). The work by Jorio et al. was done with a rAcMNPV virus but we believe BV clumping is even more problematic for HearNPV BV. Due to the study of Jorio et al we also store OB at 4 or −80 °C. OB are stable for years at 4 °C but they are nonsterile and they also clump after storage for 2 years or more. If glycerol is added to help preserve them, then they are very hard to quantify and remove from the glycerol after a few years storage and tend to clump even more.
10. HearNPV OB counts are best done with cultures harvested at 6–7 days postinfection (dpi). For synchronous infections conducted at a high MOI of 3–5 PFU/cell, OB yields typically peak by 3–4 dpi, but the OBs continue to mature and increase in size and are easier to count if harvested at 6–7 dpi.
11. OB counts are difficult to determine accurately and only by counting a large number will a reasonably accurate value be

obtained. Even then it can be subjective at times to differentiate OBs from other cellular debris of a similar size. For this reason, we also quantify OB yields via SDS gels using densitometry of the polyhedra band. When high OB/ml yields are obtained, the resulting SDS gels show a very clear large polyhedra band that is quite distinct from other protein bands on the gel.

12. While work without antibiotics is recommended, antibiotics may be supplemented to the culture medium [for example, gentamycin (Sigma) at 50 μg/ml] to maintain asepsis.
13. It is important that the culture medium used for resuspension contains 10% FBS, as FBS acts as a protease inhibitor against the action of TrypLE™ Express on cells.
14. For 75 $cm^2$ T-flasks, final culture volume is 15 ml.
15. In general, if infections are carried out for the purpose of producing working stocks infect cultures at a low MOI (i.e., MOI 0.1) to reduce the likelihood of formation of defective interfering particles. Experimental/production cultures may be infected at MOIs > 5 to ensure synchronous infection. Cytopathic effect will appear earlier in synchronous infected cultures.
16. Observe infected cultures under the microscope for the typical OrNV cytopathic effect which includes cellular hypertrophy, rounding up of infected cells, and appearance of small 'vesicle-like' structures around infected cells.
17. The need for a preconditioning period is likely to be dictated by the attachment characteristics of the individual cell line as well as the substrate material. A similar procedure for Sf-21 cells in glass roller bottles is described by Vaughn et al. (*see* ref. 42).
18. An alternative is to directly inoculate cells on microcarriers within a reduced culture medium volume of 20 ml in a spinner flask, with intermittent stirring of the culture (i.e., 3 min every 30 min) at 40 rpm for the first 12 h of culture. Following this, the culture volume is adjusted to 60 ml with fresh medium and the culture is continuously stirred at 40 rpm for the rest of the batch growth period. This method produces 91% of cell attachment to microcarriers within the first 12 h of culture.
19. OrNV does not form plaques with DSIR-HA-1179 cells, therefore the method used so far for quantification of infectious virus titer has been the $TCID_{50}$ assay based on the 50% endpoint dilution technique (*see* ref. 48).
20. The minimum number of replicates to be used with this method is 3. Accuracy of the method improves with the number of replicates. Only a noneven number of replicates can be used.

## References


1. Reid S, Chan LCL, Van Oers MM (2014) Production of entomopathoginic viruses. In: Morales-Ramos JA, Rojas MG, Shapiro-Ilan DI (eds) Mass production of beneficial organisms Invertebrates and entomopathogens. Elsevier, Amsterdam. ISBN 978-0-12-391453-8
2. Harrison R, Hoover K (2012) Baculoviruses and other occluded insect viruses. In: Vega F, Kaya H (eds) Insect pathology, 2nd edn. Elsevier, Amsterdam, pp 73–131
3. Buerger P, Hauxwell C, Murray D (2007) Nucleopolyhedrovirus introduction in Australia. Virol Sin 22:173–179
4. Ignoffo CM (1973) Development of a viral insecticide – concept to commercialization. Exp Parasitol 33:380–406
5. Black BC, Brennan LA, Dierks PM, Gard IE (1997) Commercialisation of baculoviral insecticides. In: Miller LK (ed) The baculoviruses. Plenum, New York, NY, pp 341–388
6. McIntosh AH, Ignoffo CM (1981) Replication and infectivity of the single-embedded nuclear polyhedrosis-virus, Baculovirus-heliothis, in homologous cell-lines. J Invertebr Pathol 37:258–264
7. Lua LHL, Reid S (2000) Virus morphogenesis of *Helicoverpa armigera* nucleopolyhedrovirus in *Helicoverpa zea* serum-free suspension culture. J Gen Virol 81:2531–2543
8. Nguyen Q, Qi YM, Wu Y, Chan LCL, Nielsen LK, Reid S (2011) In vitro production of *Helicoverpa baculovirus* biopesticides—automated selection of insect cell clones for manufacturing and systems biology studies. J Virol Methods 175:197–205
9. Pedrini MRS, Reid S, Nielsen L, Chan LCL (2011) Kinetic characterization of the Group II *Helicoverpa armigera* nucleopolyhedrovirus propagated in suspension cell cultures: Implications for development of a biopesticides production process. Biotechnol Prog 27:614–624
10. Mena JA, Kamen AA (2011) Insect cell technology is a versatile and robust vaccine manufacturing platform. Expert Rev Vaccines 10:1063–1081
11. Almeida AF, Macedo GR, Chan LCL, Pedrini MRS (2010) Kinetic analysis of in vitro production of wild-type *Spodoptera frugiperda* nucleopolyhedrovirus. Braz Arch Biol Technol 53:285–291
12. Micheloud GA, Gioria VV, Eberhardtb I, Visnovsky G, Claus J (2011) Production of the *Anticarsia gemmatalis* multiple nucleopolyhedrovirus in serum-free suspension cultures of the saUFL-AG-286 cell line in stirred reactor and airlift reactor. J Virol Methods 178:106–116
13. Micheloud GA, Gioria VV, Perez G, Claus JD (2009) Production of occlusion bodies of *Anticarsia gemmatalis* multiple nucleopolyhedrovirus in serum-free suspension cultures of the saUFL-AG-286 cell line: influence of infection conditions and statistical optimization. J Virol Methods 162:258–266
14. Crawford AM, Zelazny B, Alfiler RA (1986) Genotypic variation in geographical isolates of *Oryctes baculovirus*. J Gen Virol 67:949–952
15. Zelazny B, Lolong A, Crawford AM (1990) Introduction and field comparison of baculovirus strains against *Oryctes rhinoceros* (Coleoptera: Scarabaeidae) in the Maldives. Environ Entomol 19:1115–1121
16. Crawford AM (1982) A coleopteran cell line derived from *Heteronychus arator* (Coleoptera: Scarabaeidae). In Vitro 18:813–816
17. Jackson TA, Crawford AM, Glare TR (2005) *Oryctes virus*—time for a new look at a useful biocontrol agent. J Invertebr Pathol 89:91–94
18. Bedford GO (2014) Advances in the control of rhinoceros beetle, *Oryctes rhinoceros*. In oil palm—review article. J Oil Palm Res 26:183–194
19. Manjeri GR, Muhamad R, Tan SG (2014) *Oryctes rhinoceros* beetles, an oil palm pest in Malaysia. Ann Res Rev Biol 4:3429–3439
20. Portner R (2014) Animal cell biotechnology, 3rd edn. Humana, New York, NY. ISBN 978-1-62703-732-7
21. Murhammer DW (2007) Baculovirus and insect cell expression protocols, 2nd edn. Humana, Totowa, NJ. ISBN 978-1-59745-457-5
22. Drugmand J-C, Schneider Y-J, Agathos SN (2012) Insect cells as factories for biomanufacturing. Biotechnol Adv 30:1140–1157
23. Felberbaum RS (2015) The baculovirus expression vector system: a commercial manufacturing platform for viral vaccines and gene therapy vectors. Biotechnol J 10:702–714
24. Meghrous J, Mahmoud W, Jacob D, Chubet R, Cox M, Kamen AA (2010) Development of a simple and high-yielding fed-batch process for the production of influenza vaccines. Vaccine 28:309–316
25. Chan LCL, Greenfield PF, Reid S (1998) Optimising fed-batch production of recombinant proteins using the baculovirus expression vector system. Biotech Bioeng 59:178–188
26. Matindoost L, Hu H, Chan LCL, Nielsen LK, Reid S (2014) The effect of cell line, phylogenetics and medium on baculovirus budded virus yield and quality. Arch Virol 159:91–102
27. Tran TTB, Dietmair S, Chan LCL, Huynh HT, Nielsen LK, Reid S (2012) Development of quenching and washing protocols for

quantitative intracellular metabolite analysis of uninfected and baculovirus-infected insectcells. Methods 56:396–407

28. Nguyen Q, Nielsen LK, Reid S (2013) Genome scale transcriptomics of baculovirus-insect interactions. Viruses 5:2721–2747. doi:10.3390/v5112721
29. Monteiro F, Carinhas N, Carrondo MJ, Bernal V, Alves PM (2012) Toward system-level understanding of baculovirus-host cell interactions: from molecular fundamental studies to large-scale proteomics approaches. Front Microbiol 3:391
30. Chen YR, Zhong S, Fei Z, Gao S, Zhang S, Li Z, Wang P, Blissard GW (2014) Transcriptome responses of the host *Trichoplusia ni* to infection by the baculovirus *Autographa californica* multiple nucleopolyhedrovirus. J Virol 88:13781–13797
31. Visnovsky G, Claus J (1994) Influence of the time and multiplicity of infection on the batch production of *Anticarsia gemmatalis* nuclear polyhedrosis virus in lepidopteran insect cell cultures. Adv Bioproc Eng 1994:123–128
32. Pushparajan C, Claus JD, Marshall SDG, Visnovsky G (2013) Characterization of growth and *Oryctes rhinoceros* nudivirus production in attached cultures of the DSIR-HA-1179 coleopteran insect cell line. Cytotechnology 65:1003–1016
33. Nielsen LK, Smyth GK, Greenfield PF (1991) Hemacytometer cell count distributions-implications of non-poisson behavior. Biotechnol Prog 7:560–563
34. Pijlman GP, Van Schijndel JE, Vlak JM (2003) Spontaneous excision of BAC vector sequences from bacmid-derived baculovirus expression vectors upon passage in insect cells. J Gen Virol 84:2669–2678
35. Lua LHL, Pedrini MRS, Reid S, Robertson A, Tribe DE (2002) Phenotypic and genotypic analysis of *Helicoverpa armigera* nucleopolyhedrovirus serially passaged in cell culture. J Gen Virol 83:945–955
36. Lynn DE (1994) Enhanced infectivity of occluded virions of the gypsy-moth nuclear polyhedrosis-virus for cell-cultures. J Invertebr Pathol 63:268–274
37. Nielsen J, Villadsen J, Liden G (2003) Bioreaction engineering principles. Kluwer, New York, NY
38. Zhu H, Nienow AW, Bujalski W, Simmons MJH (2009) Mixing studies in a model aerated bioreactor equipped with an up- or a down-pumping 'Elephant Ear' agitator: power, hold-up and aerated flow field measurements. Chem Eng Res Des 87:307–317
39. Visnovsky G, Claus J, Merchuk JC (2003) Airlift reactors as a tool for insect cells and baculovirus mass production. LA Appl Res 33:117–121
40. Louis KS, Seigel AC (2011) Cell viability analysis using Trypan blue: manual and automated methods. Mammalian cell viability. Methods Mol Biol 740:7–12
41. Phillips HJ (1973) Dye exclusion tests for cell viability. In: Kruse PF, Patterson MK (eds) Tissue culture. Academic, London, pp 406–408
42. Vaughn JL (1976) The production of nuclear polyhedrosis viruses in large-volume cell cultures. J Invertebr Pathol 28:233–237
43. Pushparajan C (2015) Development and optimization of an in vitro process for the production of *Oryctes nudivirus* in insect cell cultures. Ph.D. thesis, University of Canterbury, Christchurch
44. Ikonomou L, Drugmand J-C, Bastin G, Schneider Y-J, Agathos SN (2002) Microcarrier culture of lepidopteran cell lines: implications for growth and recombinant protein production. Biotechnol Prog 18:1345–1355
45. Lazar A, Silberstein L, Reuveny S, Mizrahi A (1987) Microcarriers as a culturing system of insect cells and insect viruses. Dev Biol Stand 66:315–323
46. Sandford KK, Earle WR, Evans JE, Waltz HK, Shannon JE (1951) The measurement of proliferation in tissue cultures by enumeration of cell nuclei. J Natl Cancer Inst 11:773–795
47. Van Wezel AL (1973) Microcarrier cultures of animal cells. In: Kruse PF, Patterson MK (eds) Tissue culture: methods and applications. Academic, New York, NY, p 372
48. Reed LJ, Muench H (1938) A simple method of estimating 50% endpoints. Am J Epidemiol 27:493–497
49. Huynh HT, Tran TTB, Chan LCL, Nielsen LK, Reid S (2013) Decline in baculovirus-expressed recombinant protein production with increasing cell density is strongly correlated to impairment of virus replication and mRNA expression. Appl Microbiol Biotech 97:5245–5257
50. Huynh HT, Chan LCL, Tran TTB, Nielsen LK, Reid S (2012) Improving the robustness of a low-cost insect cell medium for baculovirus biopesticides production, via hydrolysate streamlining using a tube bioreactor-based statistical optimization routine. Biotechnol Prog 28:788–802
51. Agathos SN (2007) Development of serum free media for lepidopteran insect cell lines. In: Murhammer DW (ed) Baculovirus and insect cell expression protocols, 2nd edn. Humana, Totowa, NJ. ISBN 978-1-59745-457-5
52. Matindoost L, Chan LCL, Qi YM, Nielsen LK, Reid S (2012) Suspension culture titration: a simple method for measuring baculovirus titers. J Virol Methods 183:201–209
53. Jorio H, Tran R, Kamen A (2006) Stability of serum-free and purified baculovirus stocks under various storage conditions. Biotechnol Prog 22:319–325

# Part VI

## Mass Production and Formulations: Nematodes

# Chapter 10

# Formulation of Nematodes


## Arne Peters



## Abstract

The enduring stages of entomopathogenic nematodes of the genera *Steinernema* and *Heterorhabditis* are infective juveniles, which require a high humidity and sufficient ventilation for survival. Formulations must account for these requirements. Nematodes may be formulated inside the insects in which they reproduced or they need to be cleaned and mixed with a suitable binder to maintain humidity but allowing for gas exchange. Another method for formulation is the encapsulation in beads of Ca-alginate. Generic procedures for these formulation techniques are described.



**Key words** *Steinernema*, *Heterorhabditis*, Alginate, Wettable powder, Formulation


## 1 Introduction

In the pharmaceutical or agrochemistry industry, the art of formulation covers techniques and recipes that improve the targeted delivery of drugs or agrochemicals to where they are supposed to act. With biological control agents the first problem formulations are supposed to solve is mostly to keep the ingredients active over a suitable period of time. This is particularly true for insecticidal nematodes. These powerful insect control agents have a number of requirements which make their formulation difficult.

Unlike microbial biocontrol agents that are delivered as metabolically arrested fungal or bacteria spores or as insecticidal proteins, the enduring stage of entomopathogenic nematodes is a small worm with an active metabolism. It does not take up any nourishment but lives on stored lipid and carbohydrate reserves. It needs high humidity and oxygen to survive. On the other hand, it needs to get rid of gaseous end-products like carbon dioxide and ammonia. Nematodes cannot be frozen and most species will not survive temperatures above 35 °C. On the other hand, their metabolic activity produces heat which needs to be led away. To avoid the growth of contaminating bacteria, fungi, or protozoa, nematodes are often stored in the fridge at <4 °C. However, there are some species which store


Travis R. Glare and Maria E. Moran-Diez (eds.), *Microbial-Based Biopesticides: Methods and Protocols*, Methods in Molecular Biology, vol. 1477, DOI 10.1007/978-1-4939-6367-6_10, © Springer Science+Business Media New York 2016

much better at higher temperatures like *Heterorhabditis bacteriophora* (7 °C) or *Heterorhabditis indica* (15 °C) [1]. The temperature requirements of the different species vary and need to be established for every strain.

The high humidity requirements of the nematodes will enable microorganisms to grow inside the formulated product, which in turn, consume oxygen; produce ammonia, carbon dioxide, and hydrogen sulfide; cause pH shifts or even produce toxic metabolites. Fungal mycelium in the formulated product jeopardizes the dispersal in water. An obvious consequence of this is that any formulation ingredients supporting microbial or fungal growth like starch, sugar, or milk powder cannot be used for nematode formulation. A "Pesta" formulation based on such ingredients only had substantial shelf life, if formaldehyde was added [2]. Nematodes need to be washed thoroughly before being formulated to minimize the availability of dead organic matter in the package. Still, any dead nematode represents a food source for microorganisms.

The use of antibiotic compounds in the formulation may reduce the growth of fungi and bacteria but need to be selected carefully to avoid negative impacts on the nematodes and their symbiotic bacteria. Moreover, if nematode products are to be used regularly in organic horticulture, antibiotic compounds or fungicides are not allowed. In a specific case, organic herb producers in Germany identified nematode products for sciarid control as the source for residues of the fungicide Ortiva (Azoxystrobin) in their crop (J. Wienberg, pers. comm.).

The most commonly found formulation on the market is a moist dispersible powder with various silica powders as a binder. Besides there are formulations on sponge slabs, or on superadsorbant gels (mostly polyacrylate or polyacrylamide). Alginate is a popular polymerizing formulation agent for pharmaceutical, cosmetic, and nutritional ingredients and has also been employed for formulating nematodes. The polymerized alginate is strong enough to immobilize the infective juveniles (IJs). In the product exhibit, IJs of *Steinernema carpocapsae* were immobilized in sheets of Ca-alginate spread over plastic screens [3]. The nematodes had to be extracted from the alginate by adding citric acid as an activator. The principle of immobilizing nematodes in alginate is the replacement of Na+ in Na-alginate with a bivalent cation, usually $Ca^{2+}$, which triggers a crosslinking reaction resulting in a hard shell. By adding citric acid, the $Ca^{2+}$ is removed, the alginate dissolves, and nematodes can escape from the matrix. This process takes time, especially with cold temperatures which is one of the reasons why this formulation was not successful on the market.

The reversible metabolic arrestment by removing water from living organisms is called anhydrobiosis. Most entomopathogenic species do not survive low humidity but there are a few species which are more amenable to anhydrobiosis than others like *S. carpocapsae* and *S. abbasi*. Biosys has sold *S. carpocapsae* in a wettable granule where the nematodes were metabolically arrested [4] by anhydrobiosis. Oxygen consumption was decreased from 1.7 to 0.5 nl $O_2$/IJ/day after 3 days storage time. The shelf life of anhydrobiotic nematodes was superior to other formulations. However, the moisture of the granules still had to be high and allowed the growth of fungi on the surface of the granules, requiring the addition of synthetic antimicrobial compounds [5]. Moreover, the recovery of the metabolically arrested nematodes took too long. A rehydration time of 48 h was needed before the nematodes regained their full activity [6]. When applied in the field, immobile nematodes are easily inactivated by UV light or desiccation.

A prolonged shelf live by anhydrobiosis has so far only been shown for *S. carpocapsae*. Partial anhydrobiosis in *H. bacteriophora* was substantially improved by a preadaptation at a *W*-value of 0.96 [7]. The desiccation tolerance of preadapted *H. bacteriophora* can be increased by genetic selection [8] and even be stabilized by producing homozygous lines while keeping the selection pressure [9] but there seem to be a trade-off reducing the virulence and the storage stability of the selected desiccation tolerant lines.

Nematode IJs may survive harsh environmental conditions inside insect cadavers [10, 11]. Moreover, antibiotic and antifungal compounds produced by the symbiotic bacteria protect the cadaver from microbial decay [12, 13]. For small-scale applications, the most logical way of formulating nematodes is hence to leave them inside the cadaver. The IJs will stay inside the cadaver and only emerge if ambient conditions are favorable. There is evidence that nematodes emerging from infected cadavers are more infective than nematodes which were previously separated from cadavers [14].

Nematodes do not fall in the category of agrochemicals and no ingenious formulation technique will ever make them fit into the agrochemical logistic paradigm requiring a shelf life of at least 2 years at fluctuating ambient temperatures. Despite this restriction the application of nematodes against insect pests is increasing rapidly. Tailored formulations for specific applications will help to widen the use even further. This chapter provides an overview of different techniques used in formulation of nematodes and a description of some of these methodologies.

## 2 Materials

### 2.1 Production of Nematodes on *Galleria mellonella*

Larvae of the greater wax moth, *G. mellonella* can be bought as fish bait in many countries of the world. They can also easily be reared (*see* ref. 15). Last instar larvae are usually used to propagate entomopathogenic nematodes.

***2.2 Formulation Outside Insect Cadavers***

A Buchner funnel with a vacuum pump is valuable to concentrate nematodes and to remove excess water and debris. Use a Whatman No. 1 filter for holding back IJs.

***2.3 Establishing Nematode Concentration***

For nematode counting, a dissecting microscope with at least 40-fold magnification is needed, preferably with transmitting light. Accurate laboratory pipettes and a stirrer are useful to prevent sedimentation of nematodes.

***2.4 Choosing a Binder for Wettable Powder Formulations***

An accurate balance with a sensitivity of at least 1 mg is needed for monitoring water loss. A water activity meter (e.g., Aqualab from Decagon Devices, USA) is needed to establish the water activity.

***2.5 Producing Nematode-Containing Alginate Beads***

A syringe (1–10 ml) with a fine needle or a peristaltic pump with low speed is employed to create droplets. A magnetic stirrer keeps the polymerizing agent moving and a sieve (200 m mesh size) is needed to separate the beads from the polymerizing agent.

***2.6 Produce a Matrix for a Nematode Bait Station***

Polyacrylate can be obtained from Evonik under the product name Favor SXM.

***2.7 Packaging***

The temperature and humidity in the packages can be monitored with i-button dataloggers (′Thermochron' or 'Hygrochron') available from Maxim Integrated Inc. (USA).

***2.8 Evaluating Formulations***

A dissecting microscope, preferably with transmission light is needed for nematode counting. A magnification of 60- to 100-fold is sufficient for differentiating living and dead nematodes. Larvae of the mealworm (*Tenebrio molitor*) can be obtained in pet shops.

## 3 Methods

***3.1 Production of Infected Cadavers of the Greater Wax Moth, Galleria mellonella (See Refs. 11, 16)***

1. First infect insect cadavers with nematodes by placing IJs together with the insect on moist sand (8 % water content) or moist filter paper. Dead larvae may be formulated 4 days later.
2. Dip cadavers in an aqueous suspension of starch (1 %) for 3 s.
3. Subsequently roll the moist cadaver in Clay (calcium silicate) until no more clay powder adheres.
4. Cadavers are now protected from desiccation and will not stick to each other.
5. Cadavers should be used freshly (within 2 months) and placed close to the host in the soil or to another moist environment.

***3.2 Formulation Outside Insect Cadavers***

Any formulation outside insect cadavers requires a clean suspension of IJs. Remaining of the growing medium, nematode feces, other nematode stages, or fragments thereof as well as dead IJs should be removed carefully. There are different techniques available and IJs can withstand most of them.

### 3.3 Separating Living from Dead Nematodes

The separation of dead and living infective IJs is the most difficult job (*see* **Note 1**). It can be achieved by letting the IJs migrate down a White trap (*see* ref. 15):

1. Concentrate nematodes by vacuum on a filter paper on a Buchner funnel.
2. Place filter paper with nematodes on a bigger filter paper on a Petri dish lid placed in a bigger Petri dish filled with water to a level half the height of the lid inside (*see* Fig. 1).
3. Let the overlapping filter paper submerge in the water.
4. Living juveniles will actively move around on the filter paper and drop to the water while dead juveniles will remain on the filter paper. The migration will take at least 4 h at 20–25 °C. It can be done overnight.
5. Entomopathogenic nematodes are usually stored in clean tap water before being formulated (*see* **Note 2**).

### 3.4 Establishing Nematode Concentration

A first step in nematode formulation is the concentration of nematodes, i.e., removing excess liquid to get a nematode paste with 1–3 million nematodes per g. This can be done by filtration on a Buchner funnel on filter paper or by sieving the nematodes through a 20 μm nylon cloth. The number of nematodes per g needs then to be established:

1. Disperse 0.5 g in 1000 ml of water.

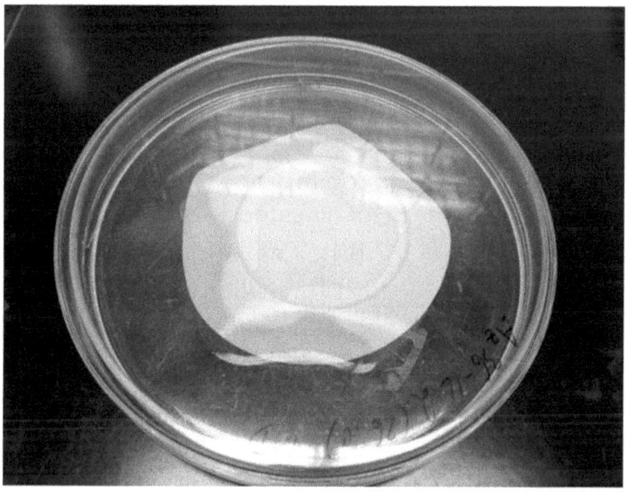

**Fig. 1** Arrangement to separate living from dead infective juveniles ("White trap") [15]. The concentrated nematodes would be placed to the center of the filter paper

2. Agitate the suspension well and then take 5 × 20 μl droplets on a Petri dish for counting under a dissecting microscope with transmission light.
3. The nematode suspension must be kept agitated during sampling, since nematodes settle quickly (*see* ref. 17).
4. Theamountofnematodespergofpastecanthenbecalculated(e.g., 100 nematodes in 5 droplets × 1000 ml × 1000 ml/100 ml/ 0.5 g = 2e6 nematodes/g).

### 3.5 Choosing a Binder for Wettable Powder Formulations

Lack of porosity in the nematode paste will block quick gas exchange and nematodes would suffocate quickly. Excess water in the paste must therefore be bound to a suitable substance. This can either be a piece of sponge or filter paper or a fine powder, resulting in a fully dispersible formulation. To find out, how much water needs to be bound to the formulation additive, the following procedure can be applied:

1. Produce nematode paste as described in Subheading 3.1 and establish nematode concentration *N* [million/g].
2. Put nematode paste on a preweighed filter paper to a layer of <1 mm and record weight *P*1 [g].
3. Let the nematodes dry at ambient temperature probably accelerated by a light airflow under a laminar flow cabinet.
4. Observe nematodes under a dissecting microscope in regular intervals (at least every 30 min). Infective juveniles will move around and form wool-like structure when most of the water is evaporated. As evaporation continues, they will continue to move in the wool until the water content gets critically low and they stop moving.
5. The weight of the nematode paste (*P*2 [g]) should be recorded again at this transition point. The difference of the weight in the beginning and the weight at this transition point gives the amount of water that should be bound to the binder in the formulation.
6. You may now calculate the water to be bound (*W*) per million of nematodes as:
   $W[\text{g} / \text{million}] = (P1 - P2) / N$, where *N* = Nematode number [in million) on the filter paper.
7. Suitable binder substances can now be screened according to their water binding capacities and to the dispersibility of the wet binder. Remember that the function of the binder is to remove excessive water from between the nematodes, to give the formulation a porous structure, to optimize gas exchange, and to keep the water activity in the package at about 0.98 (*see* **Note 3**). Water activity curves for two arbitrarily chosen binders are shown in Fig. 2.

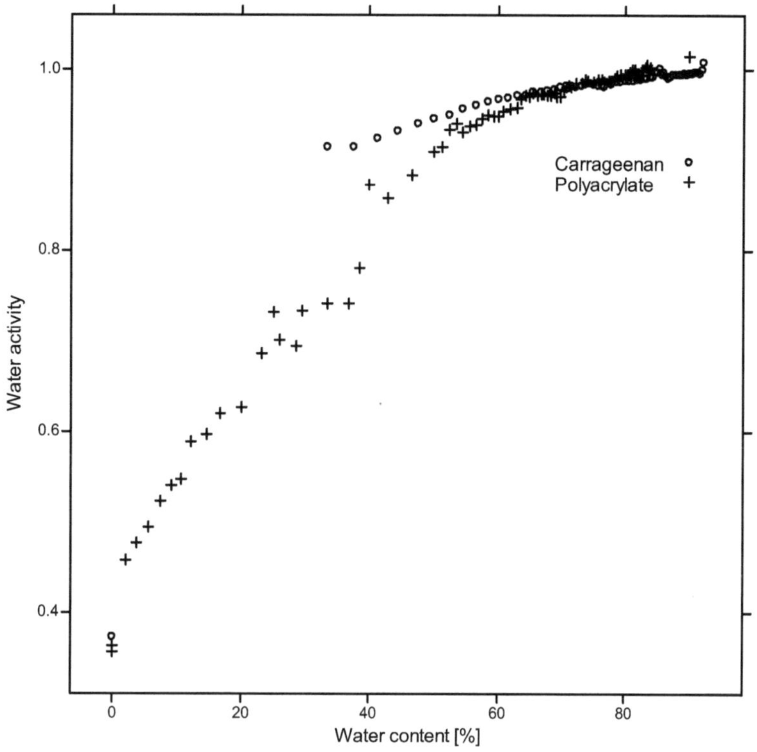


**Fig. 2** Relation of water content and a *W*-value for two potential binders for formulating nematodes. Thickener carrageenan and super-adsorbant gel (Polyacrylate)

### 3.6 Slow Release Formulations

Nematodes are still mainly applied as a drench. The formulation must hence be immediately dispersible in water. There are several pest insects, which could nicely be controlled with slow release sowable granules. Several techniques may be employed. Nematodes may be produced in insects, which are subsequently dried and applied to the soil (*see* Subheading 3.1). By choosing the right size of the insect larvae and a suitable coating technique, these nematode-infested insect cadavers may even be applied with conventional machinery (*see* ref. 14). Alginate beads (*see* Subheading 3.6) may also be suitable for slow release granules but no field data are available.

### 3.7 Producing Nematode-Containing Alginate Beads (See Fig. 3)

Alginate beads protect nematodes from desiccation and have therefore potential as bait formulations applied to the plant foliage (*see* ref. 19, 20). In the attempt to develop a slow release granule which could be sown into the soil, alginate formulations in bead form have again gained interest (*see* ref. 21). Alginate beads can be formed by letting droplets of a nematode suspension with 2 % Na-alginate fall into a suspension containing $Ca^{2+}$ (e.g., 20 mM $CaCl_2$). Reverse capsules may also be formed, i.e., the Ca ions are added to the nematode suspension and dropped into a sodium alginate suspension. This will result in a capsule with nematodes in water whereas

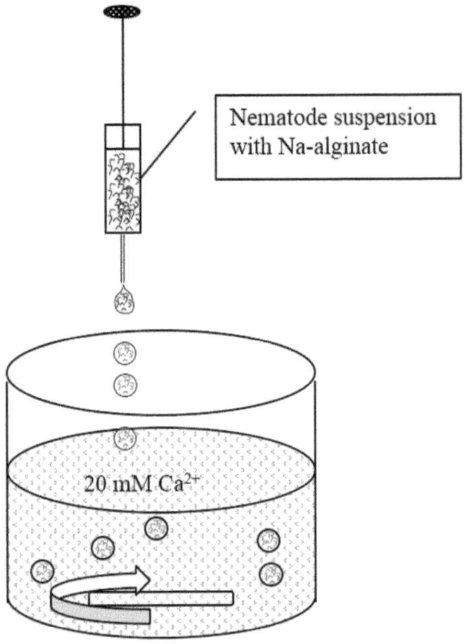


**Fig. 3** Schematic drawing for procedure to produce nematode-containing alginate beads

nematodes would reside in an alginate matrix in the first case. The important difference is that nematodes cannot push through the wall of the capsule if trapped in a liquid cavity inside the capsule while they can readily move within an alginate matrix while it has not dried out. More active nematode species like *H. bacteriophora* will therefore escape from alginate beads while more lazy species like *S. carpocapsae* would remain inside.

1. Concentrate a clean IJ suspension to the desired density (*see* Subheading 3.1) and add 2 % food-grade Na-alginate. Dispensing the alginate in an equal amount of ethanol will mitigate clumping of alginate when mixed into the nematode suspension.
2. Prepare 20 mM $CaCl_2$ polymerizing solution in distilled water in a glass beaker with a magnetic stirrer.
3. Let nematode suspension drop into the constantly stirred polymerizing suspension. Use a syringe or a thin glass tube and a pulsing pump to form droplets.
4. Collect alginate beads from polymerizing solution over a sieve.
5. Dry beads gently in air stream.
6. The strength of the capsule wall can be controlled by altering alginate and $Ca^{2+}$ concentrations and the polymerization temperature.
7. A centrifugal apparatus for producing microcapsule of highly viscous alginate suspension is presented in (*see* ref. 22).

### 3.8 Produce a Matrix for a Nematode Bait Station

In bait stations, nematode formulations must keep the moisture and facilitate nematode movement. An example is infection stations for cockroaches (*see* ref. 23). It was shown that *S. carpocapsae* is infecting cockroaches primarily via the stigmata (*see* ref. 24). The IJs are standing on their tails and move their head in a circle, they may even jump (*see* ref. 25) to get access to the host insect. On a plain agar surface, IJs are unable to lift the front part of the body and may hence fail to attach to the insect bodies surface. A few sand particles on the agar surface mitigate that problem (*see* ref. 26). A suitable matrix for an infection station should take this into account:

1. Concentrate a clean IJ suspension to the desired density (*see* Subheading 3.4).
2. Add 6% (w/w) polyacrylate (e.g., Stokosorb 500 Evonik) as a thickener.
3. Add 12% (w/w) sand (0.1–0.4 mm grain size) and mix gently to a homogenous paste.
4. Place this gel in suitable infection stations to ensure contact of waving nematodes with the most sensitive part of the insect. With cockroaches, the sides above the legs, where stigmata are located was identified as the most vulnerable site and the gel was consequently pressed to the walls of cable channels which size is chosen to fit the width of the cockroach species. A similar gel is used to control adult black vine weevils (*Otiorhynchus sulcatus*) but here the gel is placed on the ceiling of the infection chamber to enable the nematodes to attach to and move underneath the elytra (see e-nema's product Käferstopp on www.e-nema.de).

### 3.9 Packaging

At least two conflicting functions need to be fulfilled with the packaging. The humidity should be kept constant inside the package but gas exchange should be allowed for. Different package types may be used: (1) Plastic bags (20–60 μm thickness) made of polypropylene or polyethylene. Holes of <100 μm will usually keep most of the nematodes inside while allowing for air exchange. (2) Plastic trays or containers sealed with porous plastic foil or with lids which are not closing airtight.

High concentrations of IJs will produce heat which needs to be led off. In large packages, when stacked on top of each other, nematodes will kill themselves by overheating. This phenomenon has been observed especially with *Steinernema feltiae* and it is a self-reinforcing process. Increasing temperatures will speed up the nematodes metabolism and create more heat. It is therefore essential to curb the nematodes' metabolism by packing them under refrigerated conditions and by adding cooling elements to the package. Insulation of the package will retard the thawing of the cooling elements. However, the insulation will also act as a barrier for the heat once the cooling elements are thawed and the temperature in the package rises over the ambient temperature. Speed of shipment is therefore crucial.

### 3.10 Evaluating Formulations

A good formulation should keep the IJs alive (*see* **Note 4**) and infective. For evaluating formulations, the number of living nematodes and their infectivity needs to be assessed. In an attempt to standardize the counting and evaluation methods among nematode producers, standard protocols were collected (*see* ref. 27). A refined protocol is reproduced here.

#### 3.10.1 Assessing Number and Infectivity of Formulated Entomopathogenic Nematodes (See Note 5)

1. Open package and pour content in tap water of 10–20 °C in bucket of 15–25 cm diameter. Use 100 ml water per expected million nematodes.
2. Stir suspension vigorously for 1 min and keep agitated by bubbling air inside from a tube leading to the bottom of the bucket.
3. Take 3 × 100 μl samples in three clean test tubes filled with 4.9 ml tap water. Rinse pipette tip once.
4. Mix tube by shaking and immediately (<1 s) after shaking take 5 × 100 μl from each tube and place them on clean petri dishes.
5. After having placed 15 droplets count living nematodes in droplets using a dissecting microscope with 20- to 60-fold magnification.
6. To distinguish living from dead nematodes use the following criteria (*see* Fig. 4): (Alive), movement or tail or head bent or reacting on poking with needle; (Dead), shriveled surface or gas bubbles inside body.
7. The number of living nematodes per package can now be calculated taking all dilution factors into account (*see* **Note 6**).

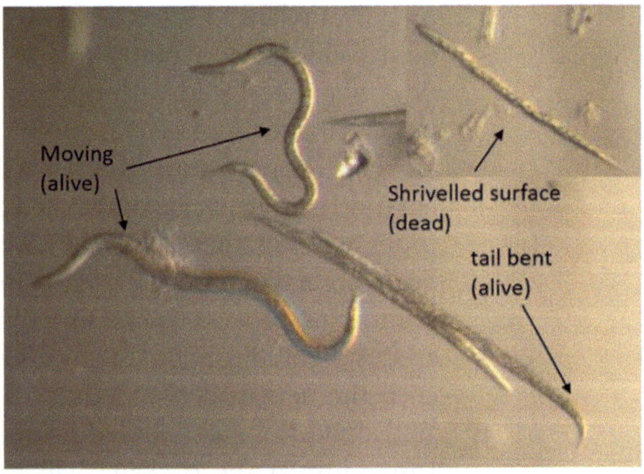


**Fig. 4** Illustration of criteria to identify living infective juveniles of entomopathogenic nematodes

*3.10.2 Infectivity Test (See Ref. 29)*

1. Prepare silver sand (0.2–0.4 μm core size) approx. 600 g for testing one batch of nematodes. Adjust to 8 % moisture content by adding tap water to dry sand.
2. Fill six plastic boxes of approx. 10 × 10 cm with moist sand to a height of approx. 2 cm.
3. Adjust nematode suspension to a density of about 30 IJs/100 μl for *H. bacteriophora*. (Values for *S. feltiae* and *S. carpocapsae* should be about 20 and 10, respectively.) Take at least ten counts of 100 μl of the final suspension and divide 30 by the mean of these counts to get the adjustment factor for the volume containing 30 nematodes.
4. Apply [100 × adjustment factor] μl multiplied by the number of insect larvae (40) to three boxes. Be sure to agitate the suspension by bubbling air or repeated stirring. Prepare three boxes with the calculated amount of nematode-free tap water as an untreated control.
5. Add 40 mealworms (*Tenebrio molitor*) to the container. Place lid on top, turn the container upside down and knock it on the bench to ensure contact of nematodes and mealworms. Incubate at 25 °C for 7 days. Punch five holes with a hot needle into the plastic lid. The lid must allow for some ventilation but the sand should not dry out during the 7 days.
6. Count dead larvae in treated and untreated boxes (*see* **Note** 7).

## 4 Notes

1. While suitable for laboratory scale, this technique cannot be scaled-up. In industry the proportion of dead IJs is sought to be minimized by culturing techniques. Counterflow techniques, like lamella separators aim at separating dead and livings IJs due their different sedimentation speed (http://www.leiblein.com/process-water/lamella-separator.html). Commercially, such sedimentation units are only available in m³ scale, but it is not too difficult to construct a lab size unit.
2. Infective juveniles of *H. bacteriophora* survive significantly better in an adapted salt solution (*see* ref. 1) at 7–10 °C and a pH of 5. Other nematodes, like *H. indica* require a much higher storage temperature. It is worth establishing the suitable concentration of salts, pH value, and temperature for new nematode isolates to be formulated and stored.
3. Although IJs are more robust than all other stages, there is a risk to damage them by shear forces during the mixing process. The shear forces may either kill the nematodes immediately or decrease survival in the package. Mixing should therefore be

done as gently as possible. Soft binders (like superabsorbent gels, polyurethane sponge, or vermiculite powder) pose a smaller risk than mineral binders with a larger particle size (attapulgite or sepiolite). To avoid any mixing stress, Bedding (*see* ref. 18) proposed a sandwich formulation where the nematode paste is wplaced between two layers of binder. If surplus of binder is used, the sandwich structure will result in a gentle desiccation during storage, similar to the granular formulation. On the other hand, nematodes in the central nematode paste layer will suffer from anoxia and/or ammonia built up. Moreover, it is difficult to scale-up this procedure. Most nematodes are currently sold in a homogenous mixture of paste and binder.

4. The IJ mortality in the formulated product over time must be evaluated by counting the concentration of living IJs rather than counting the proportion of dead IJs, since those will disappear with time and the survival would therefore be overestimated.

5. Survival in the package is temperature dependent and an accelerated shelf life testing may be adopted to evaluate a given formulation (*see* ref. 28).

6. To predict the survival, the Arrhenius equation for the speed of chemical reactions is often employed (*see* ref. 30). A more sophisticated model has been developed to predict the number of germinating fungal spores in a formulation as a function of time and temperature (*see* ref. 31). The Arrhenius equation assumes a negative exponential decrease in living units with a constant death rate (*k*). It could be shown that negative cumulative normal distributions fit the survival curves of spores much better (*see* ref. 32). Likewise, own data from *H. bacteriophora* stored in saline are better described with a negative cumulative normal distribution wthan with a negative exponential model. Interestingly a similarly good fit is achieved if a quadratic term is taken for the time in the negative exponential model (*see* Fig. 5). The reason is that in the simple negative exponential model, the death rate is constant over time whereas the death rate is increasing with time in the formula with the quadratic equation: If we replace the constant death rate ('*k*') by a death rate increasing over time ('*d* /*t*max'),

$$\mathrm{IJs} / \mathrm{ml} = N \mathrm{e}^{(-k \times t\,\dagger)}$$

Becomes,

$$\mathrm{IJs} / \mathrm{ml} = N \mathrm{e}^{(-d / t\,\mathrm{max} t\,\dagger)}$$

where: $N$ = Initial concentration of nematodes, $t$ = storage time, $k$ = constant death rate, $d$ = maximal death rate at end of storage time (set to 50 days in Fig. 5).

The better fit of the negative cumulative normal distribution or the formula with the quadratic term is therefore an indication for a death rate increasing with storage time.

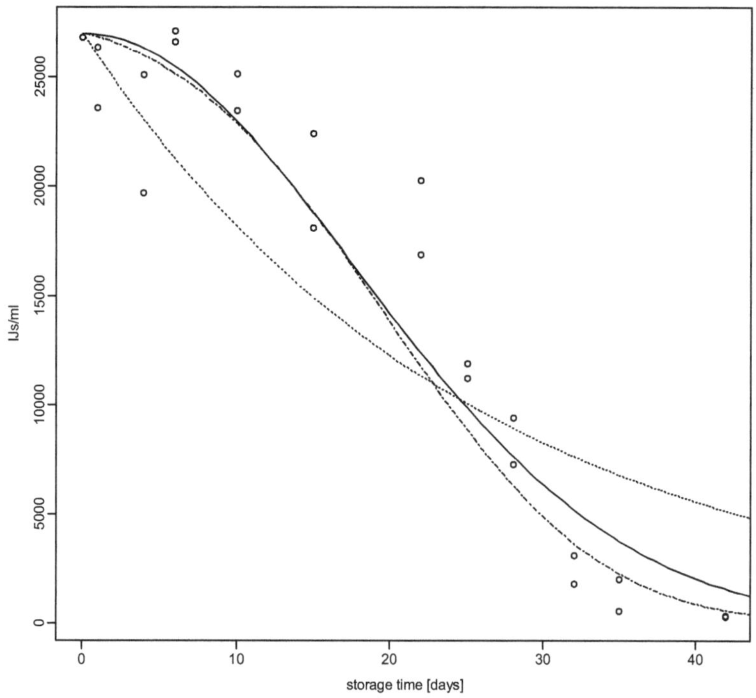


**Fig. 5** Models to fit survival data for infective juveniles of ***Heterorhabditis bacteiophora*** stored in clean water at 25 °C. The negative exponential model with a constant death rate over time gives a poor fit [IJs/ml = 29,290 $e^{(-0.041\ \text{time})}$, ***dotted line***]. If the death rate is modeled to increase with storage time by including a quadratic term for the time, the fit improves markedly [IJs/ml = 26,600 $e^{(-0.0726\ \text{time}/50\ \text{time})}$, ***solid line***] and does not deviate significantly from the cumulative negative normal distribution [Probit (IJs/ml)/28,000) = 1.83–0.092 time, ***dot-dashed line***]

7. Apart from assessing nematode mortality and infectivity there are some additional characteristics which are worth being measured in the formulation such as: water activity, oxygen concentration, $CO_2$ concentration, and ammonia concentration. The lipid content of the IJs can be measured optically or biochemically (*see* ref. 4, 33).

## References


1. Strauch O, Niemann I, Neumann A, Schmidt AJ, Peters A, Ehlers R-U (2000) Storage and formulation of the entomopathogenic nematodes *Heterorhabditis indica* and *H. bacteriophora*. BioControl 45:483–500
2. Connick WJ Jr, Nickle WR, Williams KS, Vinyard BT (1994) Granular formulations of *Steinernema carpocapsae* (strain All) (Nematoda: Rhabditida) with improved shelf life. J Nematol 26:352–359
3. Georgis R (1990) Formulation and application technology, in entomopathogenic nematodes in biological control. CRC Press, Boca Raton, FL, pp 173–191
4. Grewal PS (2000) Enhanced ambient storage stability of an entomopathogenic nematode

through anhydrobiosis. Pest Manag Sci 56:401–406
5. Grewal PS, Peters A (2005) Formulation and quality. In: Grewal PS, Ehlers RU, Shapiro-Ilan DI (eds) nematodes as biocontrol agents. CABI Publishing, Wallingford, pp 79–90
6. Baur ME, Kaya HK, Tabashnik BE (1997) Efficacy of a dehydrated steinernematid nematode against black cutworm (Lepidoptera: Noctuidae) and diamondback moth (Lepidoptera: Plutellidae). J Econ Entomol 90:1200–1206
7. Strauch O, Oestergaard J, Hollmer S, Ehlers R-U (2004) Genetic improvement of the desiccation tolerance of the entomopathogenic nematode *Heterorhabditis bacteriophora* through selective breeding. Biol Control 31:218–226
8. Anbesse S, Sumaya NH, Dorfler AV, Strauch O, Ehlers R-U (2013) Selective breeding for desiccation tolerance in liquid culture provides genetically stable inbred lines of the entomopathogenic nematode *Heterorhabditis bacteriophora*. Appl Microbiol Biotechnol 97:731–739
9. Anbesse S, Sumaya NH, Dorfler AV, Strauch O, Ehlers R-U (2013) Stabilisation of heat tolerance traits in *Heterorhabditis bacteriophora* through selective breeding and creation of inbred lines in liquid culture. BioControl 58:85–93
10. Lewis EE, Shapiro-Ilan DI (2002) Host cadavers protect entomopathogenic nematodes during freezing. J Invertebr Pathol 81:25–32
11. Shapiro-Ilan DI, Lewis EE, Behle RW, McGuire MR (2001) Formulation of entomopathogenic nematode-infected cadavers. J Invertebr Pathol 78:17–23
12. Jarosz J (1996) Do antibiotic compounds produced in vitro by *Xenorhabdus nematophilus* minimize the secondary invasion of insect carcasses by contaminating bacteria? Nematologica 42:367–377
13. Thaler J-O, Boyer-Giglio M-H, Boemare NE (1997) New antimicrobial barriers produced by *Xenorhabdus* spp. and *Photorhabdus* spp. to secure the monoxenic development of entomopathogenic nematodes. Symbiosis 22:205–215
14. Shapiro-Ilan DI, Lewis EE, Tedders WL (2003) Superior efficacy observed in entomopathogenic nematodes applied in infected-host cadavers compared with application in aqueous suspension. J Invertebr Pathol 83:270–272
15. Dutky SR, Thompson JV, Cantwell GE (1964) A technique for the mass propagation of the DD-136 nematode. J Insect Pathol 6:417–422
16. Ansari MA, Hussain MA, Moens M (2008) Formulation and application of entomopathogenic nematode-infected cadavers for control of *Hoplia philanthus* in turfgrass. Pest Manag Sci 65:367–374
17. Wright DJ, Peters A, Schroer S, Fife JP (2005) Application technology. In: Grewal PS, Shapiro-Ilan D, Ehlers R-U (eds) Nematodes as biocontrol agents. CAB International, Wallingford, pp 91–106
18. Bedding RA (1988) Storage of insecticidal nematodes. World Patent No. WO 88/08668
19. Kim YG, Lee SH, Han SC, Yu YM (2003) An edible alginate microcapsulation of entomopathogenic nematode, *Steinernema carpocapsae*. Korean J Appl Entomol 42:145–152
20. Navon A, Nagalakshmi VK, Levski S, Salame L, Glazer I (2002) Effectiveness of entomopathogenic nematodes in an alginate gel formulation against lepidopterous pests. Biocontrol Sci Techn 12:737–746
21. Kim J, Jaffuel G, Turlings TCJ (2015) Enhanced alginate capsule properties as a formulation of entomopathogenic nematodes. BioControl 60:527–535
22. Haeberle S, Naegele L, Burger R, von Stetten F, Zengerle R, Ducree J (2008) Alginate bead fabrication and encapsulation of living cells under centrifugally induced artificial gravity conditions. J Microencapsul 25:267–274
23. Appel AG, Benson EP, Ellenberger JM (1993) Laboratory and field evaluations of an entomogenous nematode (Nematoda: Steinernematidae) for German cockroach (Dictyoptera: Blattellidae) control. J Econ Entomol 86:777–784
24. Hasselmann K, Stevens AD, Heffele AD (2005) Investigations towards a biological control of cockroaches with entomopathogenic nematodes. Gesunde Pflanzen 57:169–178
25. Campbell JF, Kaya HK (2002) Variation in entomopathogenic nematode (*Steinernematidae* and *Heterorhabditidae*) infective-stage jumping behaviour. Nematology 4:471–482
26. Lewis EE, Campbell JF, Gaugler R (1997) The effects of aging on the foraging behavior of *Steinernema carpocapsae* (Rhabdita: Steinernematidae). Nematologica 43:355–362
27. Grunder JM, Ehlers R-U, Jung K (2005) Quality control of entomopathogenic nematodes. Wädenswill, Switzerland, Agroscope FAW, p 134
28. Achour M, Mtimet N, Cornelius C, Zgouli S, Mahjoub A, Thonart P, Hamdi M (2001) Application of the accelerated shelf-life testing method (ASLT) to study the survival rates of freeze-dried Lactococcus start cultures. J Chem Technol Biotechnol 76:624–628
29. Peters A (2000) Insect based assay for entomopathogenic nematode infectiousness: definitions, guidelines, problems. IOBC/WPRS bulletin 23:109–114

30. Arrhenius S (1889) Z Phys Chem 4:226–248
31. Hong TD, Ellis RH, Moore D (1997) Development of a model to predict the effect of temperature and moisture on fungal spore longevity. Ann Bot 79:121–128
32. Hong TD, Jenkins NE, Ellis RH, Moore D (1998) Limits to the negative logarithmic relationship between moisture content and longevity in conidia of *Metarhizium flavorviride*. Ann Bot 81:625–630
33. Fitters PFL, Meijer EMJ, Wright DJ, Griffin CT (1997) Estimation of lipid reserves in unstained living and dead nematodes by image analysis. J Nematol 29:160–167

# Chapter 11

# In Vivo Production of Entomopathogenic Nematodes

David I. Shapiro-Ilan, Juan A. Morales-Ramos, and M. Guadalupe Rojas

## Abstract

In nature, entomopathogenic nematodes in the genera *Heterorhabditis* and *Steinernema* are obligate parasites of insects. The nematodes are used widely as biopesticides for suppression of insect pests. More than a dozen entomopathogenic nematode species have been commercialized for use in biological control. Most nematodes intended for commercial application are produced in artificial media via solid or liquid fermentation. However, for laboratory research and small greenhouse or field trials, in vivo production of entomopathogenic nematodes is the common method of propagation. Additionally, small companies continue to produce nematodes using in vivo methods for application in niche markets. Advances in mechanization and alternative production routes (e.g., production geared toward application of nematodes in infected host cadavers) can improve efficiency and economy of scale. The objective of this chapter is to describe basic and advanced procedures for in vivo production of entomopathogenic nematodes.

**Key words** Entomopathogenic nematode, *Heterorhabditis*, In vivo, Production, *Steinernema*

## 1 Introduction

Nematodes are round worms of the phylum Nematoda. Entomopathogenic nematodes (EPNs) are insect parasites that are mutualistically associated with bacterial symbionts; the bacteria have a primary role in killing the host [1]. Although some other genera are considered to be entomopathogenic [2], in this chapter we focus exclusively on the genera *Heterorhabditis*, and *Steinernema* because they are the only entomopathogenic nematodes for which mass production methods have developed, and they are the only ones sold commercially for biocontrol purposes.

A generalized life cycle of EPNs is depicted in Fig. 1. The only free-living stage, known as the infective juvenile (IJ) or dauer stage, enters the host through natural openings (mouth, anus, spiracles), or occasionally through the insect cuticle [1]. After entering the host's hemocoel, the nematode's symbiotic bacteria are released and the host dies usually within 24–72 h. Also, once inside the insect, IJs molts, and the nematodes carry out 1–3 generations while the

Travis R. Glare and Maria E. Moran-Diez (eds.), *Microbial-Based Biopesticides: Methods and Protocols*, Methods in Molecular Biology, vol. 1477, DOI 10.1007/978-1-4939-6367-6_11, © Springer Science+Business Media New York 2016

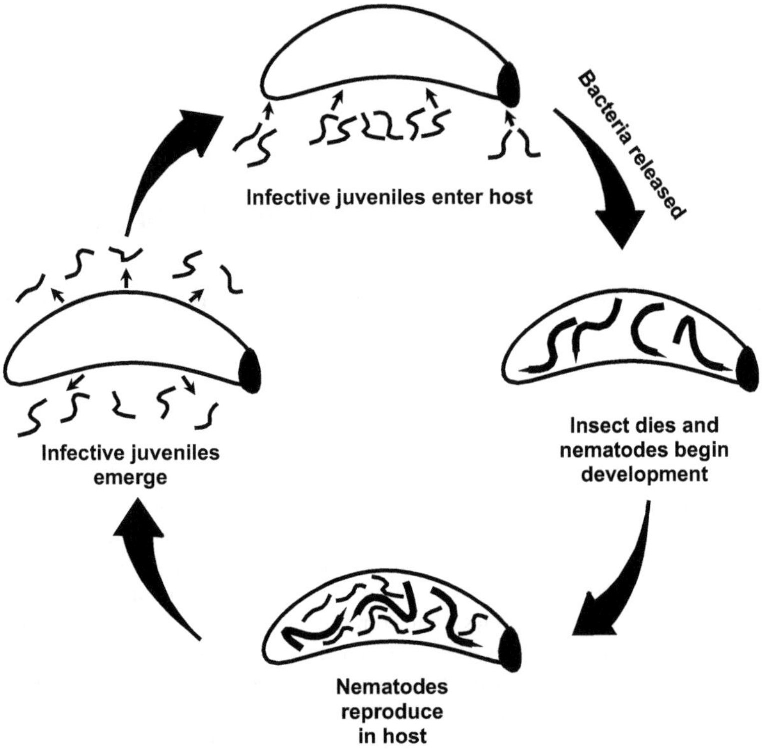


**Fig. 1** A generalized life cycle of entomopathogenic nematodes (USDA-ARS; art by B. Joyner)

bacteria also proliferate [3]. New IJs form once nutrition is depleted and then exit the insect cadaver searching for new hosts.

Entomopathogenic nematodes, which are widely known for their safety to humans and other nontarget organisms, are mass produced for purposes of biological insect control. More than 90 species of steinernematids and heterorhabditids have been described to date, and of these at least 13 species have reached commercial development [1]. The bulk of EPNs produced for biocontrol are reared in vitro (mostly in liquid culture but also to some extent in solid fermentation). However, in vivo EPN production is generally employed for purposes of laboratory research and small-scale efficacy trials in the greenhouse or field. Relative to in vitro methods, in vivo approaches have a lower economy of scale due to costs of labor and costs of insect hosts required for infection. Nonetheless, small companies continue to produce EPNs using in vivo technology for commercial application. Advances in mechanization of in vivo methodology and insect host production have led to improvements in efficiency. In this chapter, we present both basic and advanced approaches for in vivo production.

## 2 Materials

### 2.1 Basic Methods (White Trap, Culture, and Strain Maintenance)

1. Petri dishes (60–100 cm diam.) for inoculation and White traps (*see* Fig. 2).
2. Filter paper (Whatman No. 1) or similar absorbent paper.
3. Pipette (automatic or otherwise).
4. Tissue culture flasks (250–300 ml volume).
5. Insect hosts (e.g., wax worms, *Galleria mellonella*).
6. Forceps.
7. Nematode counting chamber.
8. Dissecting microscope.
9. Viable entomopathogenic nematode IJs.

### 2.2 Optimization and Scale-Up

The materials same as in the Basic Method (*see* Subheading 2.1) as well as the following:

1. Large dishes for inoculation such as 150 mm Petri dish or, for larger inoculations a metal or plastic tray up to 5000 $cm^2$.
2. If using the dunk/immersion inoculation method, a large colander and a large bowl or tank that the colander or strainer can fit into.
3. Harvest trays to mimic a large White trap and a large tub to enclose them (*see* Fig. 3), e.g., central tray may be 35 cm × 30 cm × 5 cm (depth) and holding container 90 cm × 40 cm × 12.5 cm (depth) with two harvest trays in each tub.

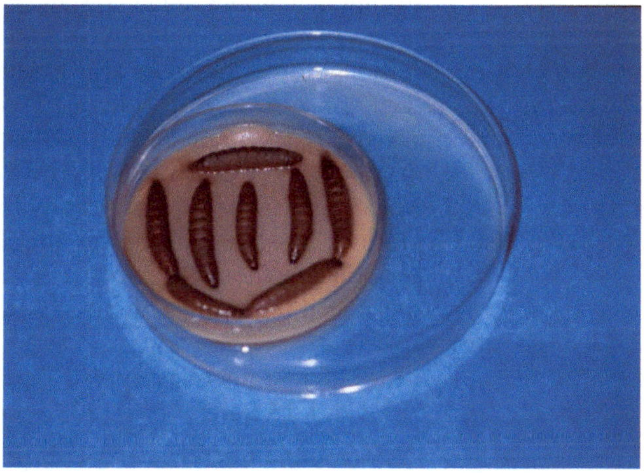

**Fig. 2** A typical White trap. The area around the inner Petri dish is filled with water below the level of dish. Nematode progeny crawl over the smaller inner dish lid into the surrounding water trap (USDA-ARS)

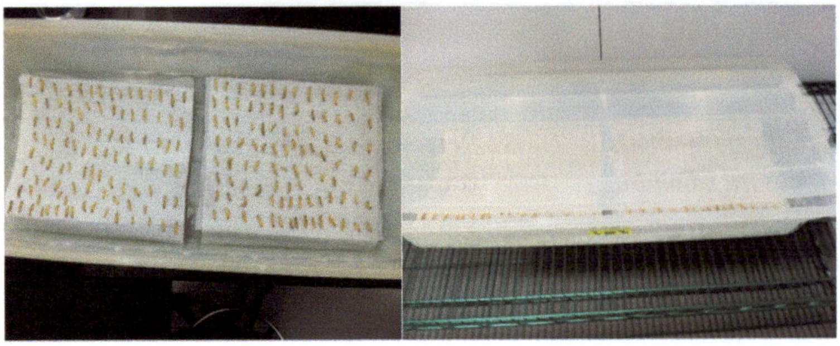

**Fig. 3** Large harvest trays with nematode-infect *Galleria mellonella*; entomopathogenic nematodes emerge and crawl into the water trap surrounding the trays (USDA-ARS; photo by S. Byrd)

4. Absorbent material for lining the inoculation and harvest trays such as paper towels.
5. Pumps and aquarium stones for aeration and concentration of IJs.
6. Holding containers or buckets for holding harvested IJs.
7. Polyurethane sponge for formulation.

### 2.3 Advanced/Automated Methods

Advanced methods described later include improved methods for rearing the insect host, *Tenebrio molitor*; automatic methods of separating optimal sizes for nematode infection; improved diets for more efficient insect rearing and for enhancing nematode virulence; mechanized methods for host inoculation; mechanized methods for formulating and packaging infected cadavers; and methods for storing nematodes in gel crystals of polyacrylamide.

#### 2.3.1 Improved Insect Production

Although other hosts are also amenable to in vivo EPN production (e.g., *G. mellonella* or others), *T. molitor* was selected as host for advanced nematode production because infected larvae tend to maintain better structural integrity than infected lepidopteran larvae, which tend to break apart easily. In addition, *T. molitor* does not produce a cocoon, which can interfere with nematode inoculation, and adults do not fly or produce scales, which eliminates the need for specialized cages and a filtration system for the air handling system.

1. Stackable fiberglass trays (Type 1) 59.37 cm long × 30.48 cm wide × 15.24 cm deep (23.375 × 12 × 6 in.) with lid.
2. Stackable fiberglass trays (Type 2) 59.37 cm long × 30.48 cm wide × 7.37 cm deep (23.375 × 12 × 3.125 in.).
3. Stackable fiberglass trays (Type 3) 65.4 cm long × 45.08 cm wide × 15.24 cm deep (25.75 × 17.75 × 6 in.) with lid and dolly (one lid and dolly for every six trays).

4. Nylon screen standard No. 20 (850 μm openings).
5. Nylon screen standard No. 35 (500 μm openings).
6. Fiberglass trays (Type 4) 52 × 39.5 × 12 cm.
7. Plastic lunch containers 2.20 × 1.50 × 0.52 cm (950 ml).
8. Food grade wheat bran.
9. Food grade dry potato squares.
10. Diet supplements (*see* Subheading 3.3.1. Diet Supplements).
11. Spray bottles.
12. Three-screen circular separator 30″ diameter (76.2 cm).
13. Screen for circular 30″ separator with rectangular openings 1.85 × 127 mm (0.073 × 0.5″).
14. Screen for 30″ circular separator with rectangular openings 1.6 × 127 mm (0.065 × 0.5″).
15. Screen for 30″ circular separator with 500 μm square openings (standard No. 35).
16. Type "Z" conveyor.
17. Vibratory feeder.
18. Hydraulic hopper lift table.
19. Seed blowers (two).
20. Diet supplements: Dry potato flour, dry chicken egg white (albumin), soy protein, peanut oil, manganese sulfate, cholesterol, reverse osmosis (RO water).
21. Planetary motion electric kitchen mixer with stainless steel bowl.
22. Kitchen spatula.
23. Electronic balance.
24. Twelve-cavity Teflon-coated cookie pans.
25. Stainless steel ice-cream scoop 2 oz (28.35 g) capacity.
26. Vacuum oven.

#### 2.3.2 Automated Inoculation and Harvest

1. For LOTEK inoculation: perforated aluminum trays, a relative humidity chamber with air vents (or humidified room), pipetting system to inoculated insect hosts (note *T. molitor* is not amenable to this method). For LOTEK harvesting: An aqueous reserve tank to distribute water, valve (controlled by time for water passage), separating screens, waste water deflector, and storage tank.
2. For automatic pipetting: Multiple stacked trays with manifold pipettes connected with hosing to a central inoculation tank. Nematodes and host insects.
3. For the Gel System: polyacrylamide gel (or other suitable gel), bags to hold the gel with a Tyvec® strip added for aeration, nylon screen (1–2 mm holes; polyester or other screen may be used as well) with a plastic lip to hold insect host cadavers.

*2.3.3 Production and Formulation of Cadavers*

1. Automatic insect cadaver packing machine.
2. Air compressor with a minimum capacity of 150 PSI.
3. Masking tape 1.5 in. wide (3.6 cm).

## 3 Methods

### *3.1 Basic Methods (White trap, Culture, and Strain Maintenance)*

Generally, in vivo approaches are based on two-dimensional systems consisting of production in trays and shelves (*see* refs. 1, 4). Most systems revolve around the concept of a White trap (*see* refs. 5, 6), which is a device used for harvesting IJs that takes advantage of the progeny IJ's natural migration away from the host cadaver upon emergence. Overall, the basic approach consists of inoculation, harvest (e.g., via White trap), concentration, and if necessary, decontamination.

1. Nematodes are inoculated into Petri dishes (or similar dish or tray) lined with filter paper or any other inert absorbent material. Inoculation is performed by applying IJs in aqueous suspension onto the absorbent material.
2. The concentration of nematodes required varies by insect host and nematode species. The last instar larvae of the greater wax moth, *Galleria mellonella*, is the most common host used because it is extremely susceptible and easy to obtain (e.g., in a number of countries the insect is produced commercially for fish bait or pet food). For most nematode species, a rate of 25–100 IJs per *G. mellonella* larva is sufficient to produce infection.
3. The concentration of nematodes can be determined by placing a known volume of stock suspension on a dish or preferably using a nematode counting chamber and then calculating the number of IJs per ml. The counts should be made under a dissecting microscope. If necessary the stock solution can be diluted for ease of counting (optimum numbers to count should fall between 30 and 200) and then the concentration can be calculated based on the dilution factor. For example, if a 100× dilution is made from the original stock solution (1 ml IJ suspension + 99 ml water) then the actual count should then be multiplied by the dilution factor (100×). For more accurate estimates, it is best to count the sample at least three times and obtain an average.
4. The total volume of IJ suspension applied to each dish or tray will vary depending on size of the tray and type of absorbent material. For example, if using a 100 mm Petri dish lined with Whatman No. 1 filter paper a total volume of 1 ml aqueous suspension can be applied whereas approximately 0.35 ml of suspension should be applied if using a 60 cm Petri dish. It is critical that a moist surface is maintained on the medium to

allow the nematodes to move and to help create the high relative humidity required. However, excess moisture (i.e., standing water) is detrimental.

5. The number of insects in the dish is also important. For example, if using a 100 mm Petri dish then ten *G. mellonella* larvae per dish is a suitable number. Thus, with the 100 mm Petri dish and ten *G. mellonella* one would apply 1 ml of EPNs at 250–1000 IJs per ml. The application of suspension can be made using any pipetting devise.
6. Once inoculated, the dishes should be incubated for 2–4 days before infected cadavers are transferred to White traps. The optimum incubation temperature will vary by nematode species (*see* ref. 1). Most nematodes can reproduce well at 25 °C, which therefore serves as a standard in most cases. However, certain species such as *Steinernema feltiae* have lower optimum temperatures (e.g., closer to 21 °C) whereas others such as *Steinernema riobrave* prefer higher temperatures such as 30 °C.
7. After incubation, the nematode-infected cadavers are transferred to White traps. Only those cadavers showing patent infections (typical signs of infection for the pathogen of interest) should be transferred. Patent infections for heterorhabditids are typically indicated by a reddish color and for steinernematids a brown or tannish coloration should appear in the infected host (*see* refs. 1, 6). Black or putrid cadavers should be removed. The White trap consists of a dish or tray on which the cadavers rest; the dish is surrounded by water, which is contained by a larger arena (*see* Fig. 2).
8. Once IJs begin to emerge the nematodes should be harvested daily until emergence ceases. Depending on nematode and species as well as incubation temperature, emergence may begin as early as 7 days postinfection and can last for 3 weeks. In *G. mellonella*, emergence is likely to begin 9–11 days postemergence and the bulk of IJs will have exited by about 16 days. Usually, the heaviest emergence occurs within 5–7 days after its initiation (and one may choose to stop harvesting at that point).
9. Harvest is achieved simply by removing the center dish of the White trap and pouring off IJs into a suitable storage vessel. Most commonly, IJs are stored in tissue culture flasks.
10. The IJs should be stored under refrigeration. For most steinernematids, a storage temperature between 4 °C and 10 °C is suitable whereas most heterorhabditids store best at temperatures between 4 °C and 13 °C (depending on species).
11. In aqueous suspension (such as in culture flasks), IJs should be kept at concentrations lower than 10,000 per ml to avoid overcrowding and oxygen deprivation. Also, the volume of suspension in the culture flask should allow for adequate oxygenation,

e.g., a maximum of 20 % of the vessel's volume should be liquid. Depending on nematode species, IJs can generally be stored in this manner for 2–6 months. However, if the IJs are to be used for experimental purposes, it is customary to only allow 2–3 weeks of storage prior to use to ensure maximum fitness of the nematodes (*see* **Notes 1–7** for assessment and maintenance of quality and fitness in IJs during in vivo production).

### 3.2 Optimization and Scale-Up

Based on the simple White trap methods indicated earlier, the process can be optimized and scaled-up to suit the needs of small field trials or cottage scale commercial ventures. Aspects that can be optimized and scaled-up include nematode species or strain as well as host species, inoculation rate and approach, host density and tray size, harvest, storage, and environmental conditions.

1. The choice of nematode species is critical and can make a huge difference in IJ yields. However, the nematode choice also depends heavily on which insect pest one may be targeting (as virulence will vary by species and strain as well). Variation in yield among nematode species is roughly inversely proportional to IJ size, yet some species simply have innately high reproductive capacities, such as *H. indica* and *S. riobrave*. For example, yields of *S. riobrave* (average IJ length = 622 mm) may exceed 300,000 IJs per insect in *G. mellonella*, whereas for a large nematode such as *S. glaseri* (average body length of IJ = 1133 mm) yields do not exceed 50,000 IJs in the same host. Certain strains (within species) also may vary substantially in yield.
2. The choice of host insect is also important. As mentioned earlier, *G. mellonella* is the most common insect host used for in vivo laboratory and commercial EPN production. Only a few EPNs exhibit relatively poor reproduction in *G. mellonella* (*S. kushidai*, *S. scapterisci*, and *S. scarabaei*). The yellow mealworm, *T. molitor*, has also been used for in vivo production extensively. A few others that have been studied for in vivo production include tobacco budworm, *Heliothis virescens*, pink bollworm, *Pectinophora gossypiella*, corn earworm, *Helicoverpa zea*, and house cricket, *Acheta domesticus*.
3. Nematode yield is generally proportional to insect host size, yet IJ yield per mg insect (within host species), and susceptibility to infection, is usually inversely proportional to host size or age. In addition to yield, ease of insect culturing and susceptibility to IJs are important factors when choosing a host. Finally, the choice of host species and nematode for in vivo production should rely on nematode yield per cost of insect, and the suitability of the nematode to the target pest.

4. Approaches for optimization of inoculation procedures can be modeled on Shapiro-Ilan et al. (*see* ref. 7). The method of inoculation can be important and may be optimized depending on host and nematode species. The goal is to expose nematodes to insect hosts and reach infection in the most efficient manner possible. Options include pipetting, applying nematodes to insect food, or dunking the hosts in a nematode suspension. Pipetting is the most common method; various types of pipettes can be used. Applying nematodes to insect diet tends to be inefficient because the food will need to be removed at a later time. However, for certain cases where infectivity is low using other methods, higher rates of infection may be achieved using the diet method.
5. The dunking or immersion method of inoculation can be very efficient. For most EPN species, dunking last instar *G. mellonella* in suspensions of IJs concentrated at 5000–7000 per ml can achieve high rates of infection. Relatively large numbers of larvae can be dunked at a time (e.g., 500–1000). The hosts are dunked into a suspension of IJs using a large strainer or colander (such as one used for straining pasta). The insects should be submerged completely for a short duration such as 1–3 s. After submersion, the insects are dumped onto a large tray containing absorbent material, most commonly paper. Note, some host–nematode combinations are not amenable to the dunking method, e.g., *T. molitor* and *H. bacteriophora* (*see* ref. 7).
6. Regardless of the inoculation method, the inoculation rate (concentration of IJs and amount applied) should be optimized for each particular host and nematode species. The goal is to reach as close to 100 % patent infections as possible because subsequent removal of noninfected or poorly infected hosts is time consuming. However, if IJ concentration is too high (higher than needed to maximize the infection rate) then the percentage of nonpatent infections may increase.
7. In addition to optimizing inoculation method and rate, the host density should be optimized for maximum efficiency. If density becomes too high the yield per insect will decrease due to overcrowding.
8. As suggested earlier, optimum inoculation parameters will vary by host and nematode species. Nonetheless, in one study (*see* ref. 7) optimum concentrations reported were 4000 IJs per ml for dunking *G. mellonella* in suspensions of *H. bacteriophora* or *S. carpocapsae*, 21,000 IJs per ml for inoculating *T. molitor* with *S. carpocapsae* using the dunking method, and 400 IJs per insect for inoculating *T. molitor* with *H. bacteriophora* using the pipette method (as dunking was not effective for this combination).

9. Furthermore, in terms of host density, optimum levels to maximize the yield per tray were 0.07–0.13 g host per cm$^2$ for *G. mellonella*, and 0.04 g host per cm$^2$ for inoculation of *T. molitor*. Ideally, inoculation parameters should be optimized for each production operation individually.
10. The size of tray used for inoculation depends on the scale desired. At a certain point the tray becomes too large and unwieldy. Some examples of scaled-up inoculations tray sizes may include 180–3000 cm$^2$. The trays can be of essentially any hard material (e.g., plastic or metal) as long as an absorbent surface such as paper can be lined within.
11. Once insect hosts are inoculated, the trays must be incubated at a suitable temperature (optimized for each nematode species or strain) and then infected hosts are transferred to harvest trays after 2–4 days.
12. Scaled-up harvest trays that mimic large White traps can be constructed in variable sizes (*see* example in Fig. 3). The infected hosts rest upon absorbent moist material on the raised harvest tray and emerging IJs migrate into a surrounding water trap. The trays should have paper or similar absorbent material around the sides to wick water. The harvest trays can be enclosed, but some aeration should be allowed (lids can be left ajar). Holes in the trays through the absorbent material may allow for increased movement of IJs into the water below. The water should not be too deep and thereby cause oxygenation problems; approximately 2–4 cm depth is suitable.
13. IJs can be harvested from the large White trap harvest trays by pouring the suspension into a bucket. Alternatively, if there are numerous trays, the IJs can be pumped into a central collection tank.
14. Throughout the process, environmental conditions should be optimized such as for temperature, aeration, and relative humidity. To minimize overcrowding effects leading to oxygen deprivation (infected cadavers can emit harmful gases such as ammonia), precautions should be taken to allow for adequate airflow. Ideally, a pass-through HEPA filter system is implemented. However, aeration should be balanced with maintaining a high humidity which is also critical for nematode productivity within the host. If humidity is kept sufficiently high in the production room (close to 100%) then it is not necessary to enclose the harvest trays.
15. Following harvest, if refrigerated storage in culture flasks is not feasible due to the large number of IJs collected, the nematodes can be kept under refrigeration in aqueous suspension in buckets or other large containers as long as the suspensions are fully aerated. Aeration can be achieved with normal pumps such as aquarium pumps and bubble stones, which come in various sizes. The nematodes can be stored in this manner at concentrations up to 50,000 IJs per ml for up to 2 months.

16. Rather than holding the IJs in large quantities of aqueous suspension, it may be preferable to concentrate and formulate the nematodes in a carrier. For commercial purposes this is considered to be a necessary step as large quantities of liquid suspension are unwieldy. Concentration can be achieved by gravity settling, but prolonged periods of settling may be detrimental due to oxygen deprivation.
17. The concentration process can be accelerated by vacuum filtration. One method is to use reverse suction on an air pump and pull water through an aquarium stone. The IJs will concentrate into a paste-like substance (appearing somewhat like peanut butter) at which time the paste can be spread onto sponge or mixed with another suitable carrier (*see* ref. 4). Centrifugation may also be used for concentrating IJs, but the costs for suitable centrifuges that would be appropriate are generally prohibitive for in vivo operations.
18. Sponge formulations can hold approximately 0.1 million IJs per $cm^2$ and may be stored for 6 months or more under refrigeration. Other formulations such as certain types of vermiculite, peat, or clay have been used to hold much higher concentrations of nematodes.

### 3.3 Advanced/Automated Methods

#### 3.3.1 Improved Insect Production

Insect production for *T. molitor* consists of three basic systems including the reproduction or adult system, the growth or larval system, and the separation system.

#### 3.3.2 Tenebrio Molitor Rearing

An environmentally controlled rearing room is required. The preferred environmental conditions are between 24 and 27 °C and a minimum of 70% RH. Lights are not required and the room may be kept dark during most of the time. Late larval stages of *T. molitor* could benefit by lower temperatures because they produce significant metabolic heat, which can increase the temperature of the rearing trays by 5–9 °C depending on the density (Morales-Ramos, unpublished).

1. *Tray modification.* Type 1 trays are modified by cutting the bottoms of the trays and replacing the bottoms with nylon screen standard No. 20 (850 μm). Three circular windows are cut in each of the longer sides and covered by nylon screen for ventilation. Type 3 trays are modified in a similar way, but the screen used to replace the bottom is standard No. 35 (500 μm). Circular windows as described earlier are cut, 6–7 in each of the longer sides and three in the shorter sides of type 3 trays and covered with nylon screen. Trays type 2 and 4 remain unmodified.
2. *Reproduction.* The adult tray system consists of one modified type 1 tray stacked on one unmodified type 2 tray (*see* Fig. 4). Tray type 1 holds the food and adult beetles, and tray type 2 collects first instars. From 250 to 300 adults (1:1 sex ratio) are introduced in each tray type 1 and provided with 400 g of wheat bran and 30 g of adult supplement. The food provided

is sufficient to maintain the adults for 2 months and additional feeding is not required. Adults must be provided with water twice a week by using a spray bottle to distribute water uniformly. The adult density range recommended here (between 8.4 and 14 adults per $dm^2$) has been determined to be optimal to minimize cannibalization of eggs by adults and to reduce adult mortality (*see* ref. 8). Female beetles oviposit and glue their eggs on the surfaces of the tray and on the food particles. Eggs hatch in approximately 8 days at 26 °C (*see* ref. 9) and first instars tend to migrate to the bottom of the tray where they fall down through the nylon screen to the collection tray (*see* Fig. 4B).

**Fig. 4** Adult tray system. (**A**) Tray type 1 (*a*) open and showing adults with food sitting on top of a tray type 2 (*b*). (**B**) Tray type 2 (*b*) open showing small food particles with first instars ready for collection

3. *Progeny collection.* First instars are collected once a week and the collection from each tray is placed in one plastic lunch box and provided with 4 g of wheat bran. Adults are maintained in production for a period of 8–9 weeks. During this period, adult females produce 85–90 % of their oviposition potential and older females will produce few eggs (*see* ref. 8). Adults older than 9 weeks are discarded and replaced by newly emerging ones.
4. *Growth.* Small larvae are maintained in the plastic lunch boxes for a period of 6 weeks and then transferred to the larval tray system (*see* Fig. 5A). The larval tray system consists of 1–5 modified type 3 trays stacked on top of one unmodified type 3 tray (*see* Fig. 5B). All six trays are placed on a dolly and the top tray is covered with a lid (*see* Fig. 5A). At the end of 6 weeks, the contents of between six and eight lunch containers (each containing the larvae collected from one adult box per week) are transferred to one of the

**Fig. 5** Larval tray system. (**A**) Stacks of modified type 3 trays sitting on top of one unmodified type 3 tray and a dolly. (**B**) Open system showing larvae with food on a modified tray (*a*) and frass collected in the unmodified tray at the bottom (*b*)

modified type 3 trays. Higher larval densities will increase development time and larval mortality reducing the productivity of the system (Morales-Ramos, unpublished). Each larval tray is provided weekly with between 400 and 500 g of wheat bran depending on the feeding rate. In addition, each tray is provided with 20–40 g of larval supplement. If the environmental conditions in the room are maintained with a relative humidity higher than 70%, larvae do not require watering. If difficulties exist in maintaining this RH level, larvae must be provided with water at least twice a week by spraying or dripping. Larvae remain in this tray system for a period ranging between 6 and 9 months depending on temperature, larval density, and humidity. Food consumption by larvae should be monitored weekly. A drastic drop in food consumption indicates that larvae are close to pupation and they are ready for separation (Morales-Ramos, unpublished).

5. *Separation*. Because *T. molitor* exhibits developmental plasticity, the larval stage goes through a variable number of instars ranging from 12 to 22 (*see* ref. 10). Variable numbers of instars result in substantial variability in the larval development time (*see* refs. 9, 11). This means that synchronization of developmental stages is not possible in *T. molitor* and therefore, larvae must be separated by sizes even if they belong to the same cohort. The process of separation has been mechanized by combining existing technologies into a novel insect separation process (*see* ref. 12) (*see* Fig. 6A). When larvae are ready to be separated, a larval stacked tray system is rolled into the separation room, which holds the separation system. The separation system (*see* Fig. 6A) consists of a vibratory feeder (a), a conveyor (b), a 3-screen circular separator (c) with four dischargers (*see* Fig. 6C), and two seed blowers (d). Trays are emptied into the vibratory feeder, which distributes larvae mixed with food in a uniform way into the conveyor (*see* Fig. 6B). The conveyor moves the larvae and food mixture into the 3-screen separator by dumping the material into the top opening. Because the food material may contain fine dust it is advisable to install an aspiration or vacuum system connected to a bell-shaped collector on the top of the separator to collect fine dust. A similar system may also be required on top of the vibratory feeder. The larvae and food mix will pass through three separation screens inside the machine.

6. *Distribution of separation portions*. The first screen has rectangular openings (1.85 × 127 mm) and the second screen has smaller rectangular openings (1.6 × 127 mm). Rectangular openings facilitate the separation of larvae by minimizing their ability to grab to the screen. The third screen is a conventional standard No. 35 screen with 500 μm square openings. Food and larvae are separated into four groups which are released from four dischargers located around the separator (*see* Fig. 6C). The first discharge group includes the largest larvae, which are

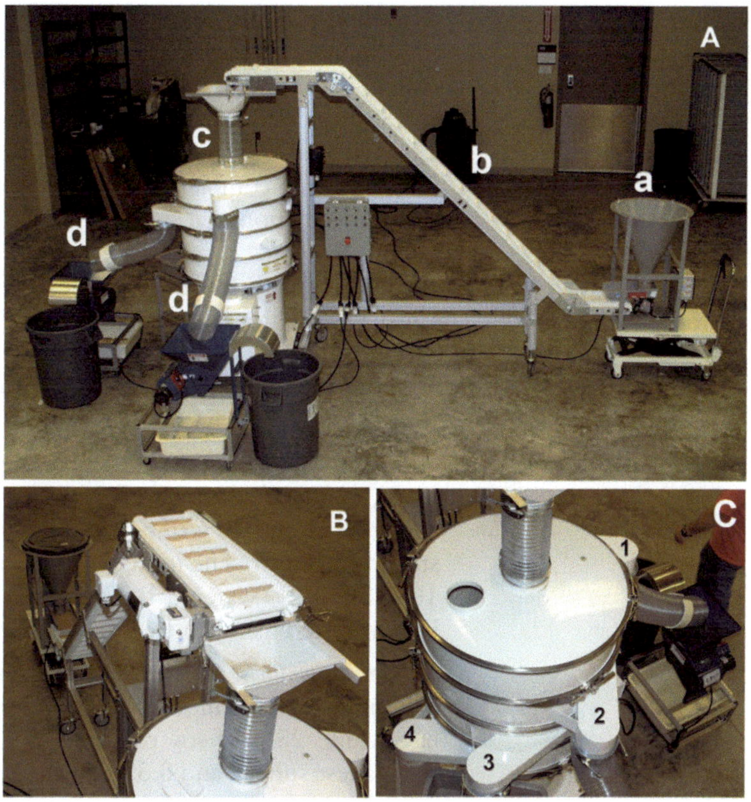


**Fig. 6** Larval separation system (**A**) consisting of a vibratory dispenser (*a*), conveyor (*b*), three-screen circular separator (*c*), and two seed blowers (*d*). (**B**) The vibratory feeder spreads larvae and food on the conveyor, which transports and dumps the mix into the separator. (**C**) The three screen divides the mix into four discharges consisting of (*1*) large size larvae, (*2*) intermediate size larvae, (*3*) small size larvae mixed with food, and (*4*) frass

suitable for nematode infection or for colony reproduction. The second discharge group includes intermedium size larvae that can be used for nematode infection, returned to the colony to continue developing, or can be sold for different purposes. The third discharge group includes smaller larvae mixed with food and this portion is returned to the colony to continue developing. Only frass particles and trash will emerge through the last discharger. Frass can be utilized as organic fertilizer.

7. *Pupae collection*. A portion of the older larvae obtained from the first separator discharger during the separation process are returned to the colony to obtain new adults for reproduction. These groups comprise larvae of 1–3 instars before pupation. Because *T. molitor* has a variable number of instars, pupations are not synchronized and can be spread out over a period of 2 months or more. Older larvae are placed in unmodified type 4 trays and monitored daily for pupation. Pupae are collected

daily and placed in an environmental chamber at 15 °C to stop development. Pupae can remain at this temperature for up to 2 weeks without suffering adverse effects. Once a week, a group of 350 of the stored pupae are selected and transferred to a clean type 4 tray lined with tissue paper and placed at rearing conditions to complete development. Resulting adults are used for reproduction.

#### 3.3.3 Diet Supplements

Two supplement formulations are used for *T. molitor* production. The first formulation is used to increase fecundity of adult beetles (adult supplement) (*see* ref. 9) and the second formulation is used to increase larval survival and nematode virulence (larval supplement) (*see* ref. 13).

1. *Adult supplement preparation.* To prepare 500 g of dry supplement, 420 g of dry potato flour, 60 g of dry egg white, 10 g of soy protein, and 10 g of peanut oil are dispensed into a stainless steel mixer bowl. The ingredients are mixed using a spatula and 1000 ml of RO water (1 l) is added to the mix. The formulation is then mixed using an electric blender for 20 min.
2. *Larval supplement preparation.* Ingredients required for 500 g of dry supplement include, 398.9 g of dry potato flour, 50 g of dry egg white, 25 g of soy protein, 25 g of peanut oil, 0.75 g of manganese sulfate, and 0.35 g of cholesterol. The ingredients are placed in a stainless steel mixer bowl and mixed using a spatula. Then 1 l of RO water is added to the blend and mixed for 20 min using an electric blender.
3. *Supplement pellet preparation.* Using an ice cream scoop (60 ml), portions of the supplements are dispensed into the cavities of a cookie pan. The cookie pans are introduced into a vacuum oven at 58 °C and exposed to a negative pressure of 1010 mbar for a period of 48 h. The dry supplement pellets should weigh approximately 27 g each. Supplements are dried in this manner to facilitate storage and lengthen shelf life. They can be stored for more than 1 year if maintained dry. Adults and larvae have no problem eating the dry supplement, which can be added directly into the rearing trays.

#### 3.3.4 Automated Inoculation and Harvest

With improved low cost and highly fit host insects in hand, the next step in automating the in vivo EPN production process is to mechanize inoculation and harvest. Any approach that reduces labor will contribute to cost efficiency.

1. Inoculation can be mechanized using a multiple or automated pipette system. Multiple plastic trays (e.g., approximately 2500 $cm^2$) with paper towel lining, or a similar absorbent material, can be stacked in a shelf-like manner. Each tray will hold host insects (*G. mellonella* or *T. molitor*) at optimized densities

(see earlier). A central aerated-bubbling tank of IJ inoculum is then pumped into multiple pipettes that cover the area of each tray. Several shelf systems can be added in parallel. The shelves can be open to the air if they are in a room with high humidity (>95 %) or they can be enclosed with an associated mechanism to add humid air (via pumping or a pool of water below). The infected insect hosts are then incubated and transferred after 2–4 days for harvesting.

2. One option for automated inoculation and harvesting is the LOTEK system (*see* ref. 14). A mechanized harvest device, LOTEK, allows for automated collection of IJs from stacked trays; the IJs are pumped to a central collection tank; unlike the White trap method, nematode migration into a water reservoir is not required. Briefly, the system consists of perforated aluminum holding trays to hold insect hosts during inoculation and harvesting. The trays can be 30 cm × 26 cm × 4 cm (*see* ref. 14) though they can also be customized as needed. Perforations in the tray should be small (e.g., 1.6 mm) to allow passage of nematodes but prevent insect hosts from passing through. The insects, such as *G. mellonella* larvae, are inoculated directly on the trays and incubated in stacked fashion. The trays are then mounted on an automated harvester with misting nozzles. The misting washes emerging nematodes downward where are they collected and concentrated into a storage tank.
3. In another approach, the "Gel System" automated harvesting is combined with a final packaging step in the same setup (*see* ref. 15). Infected host cadavers are placed on a substrate intended as the formulation carrier (such as a gel) within the final package; once IJs emerge the cadavers are removed and that the final nematode product is ready for shipment or storage (*see* Fig. 7). In one example, the carrier is moist polyacrylamide gel (1 g gel to 75 ml water), a nylon screen (hole size approximately 1–2 mm) with plastic rim holding the infected hosts is placed on top of the gel, and IJs emerge naturally into the gel; the gel and harvest screen

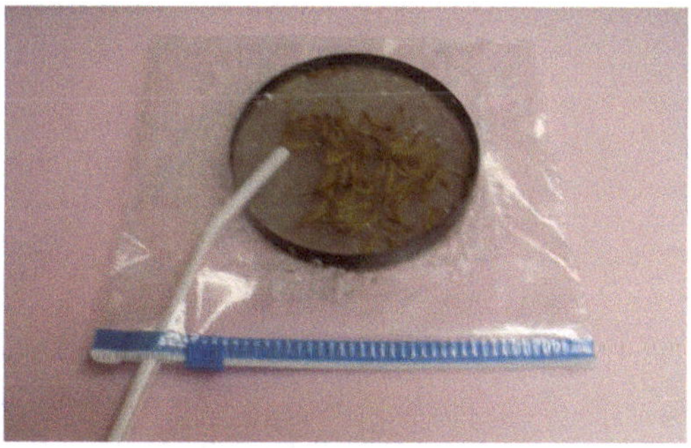

**Fig. 7** "Gel System" setup with entomopathogenic nematode infected *Tenebrio molitor* larvae on a screen on top of a polyacrylamide gel (USDA-ARS)

are enclosed in a ventilated plastic bag. Ventilation can be accomplished with a simple straw (*see* Fig. 7) which is removed once IJ emergence ceases or, a Tyvek® strip can be used to ventilate the bag efficiently (and this is less cumbersome than the straw). After the IJs have emerged the nylon screen and host cadavers are removed. The bag containing gel and IJs is then ready for storage and usage (e.g., commercial sale or experimentation). Five to ten million IJs can be stored in bags originating from 1.0 g of gel (e.g., 100 infected *T. molitor* may be suitable for this size). The size of the apparatus can be altered to suit the user's needs.

#### *3.3.5 Production and Formulation of Cadavers*

An alternate approach for in vivo production is culture and delivery of EPNs in their infected host cadavers (*see* ref. 1). In this method, nematodes are applied to the target site in their infected hosts, and pest suppression is achieved subsequently by progeny IJs that emerge from the cadavers. Production and application of nematodes in infected hosts may be more efficient than other in vivo production methods because harvest and concentration steps are removed from the process. Furthermore, certain studies have indicated that nematodes applied using the cadaver approach exhibited superior dispersal, infectivity, survival, and efficacy (*see* ref. 1).

Infected hosts are produced as described earlier except nematodes are not harvested from the cadavers; rather, the infected hosts are stored or used directly for application. Infected hosts can be produced using *G. mellonella*. However, when G. mellonella are infected with EPNs the cadavers are relatively fragile and can stick together. To overcome this issue, host cadavers can be formulated to enhance their integrity (*see* refs. 1, 16). One suitable formulation consists of dipping the cadavers in a starch suspension (1 %) and then rolling them in a clay powder (*see* ref. 16). An alternative approach is to use hard bodied insects as the host cadaver, such as *T. molitor*, which have some natural resistance to rupturing or sticking together upon handling (*see* ref. 17).

To further the ease of use, cadavers of *T. molitor* larvae infected by EPNs can be packed between two sheets of masking tape using a specially designed automatic machine (*see* ref. 18) (*see* Fig. 8A). This machine operates by pneumatic actuators and it is computer controlled.

1. The packing machine is loaded with two rolls of masking tape, connected to an air pressure source, turned on and set in standby mode. The cadaver receptacle (*see* Fig. 8B) is filled with infected *T. molitor* larvae cadavers. When the machine is activated, individual cadavers are picked up and positioned on the sticky side of a masking tape strip by a mechanical positioner (*see* Fig. 8B, b). The machine moves the tape with adhered cadavers 7 cm per cycle at the end of which it deposits a new cadaver on the tape. As the tape with adhered cadavers moves, a second strip of masking tape is placed on top of the first thereby enclosing the

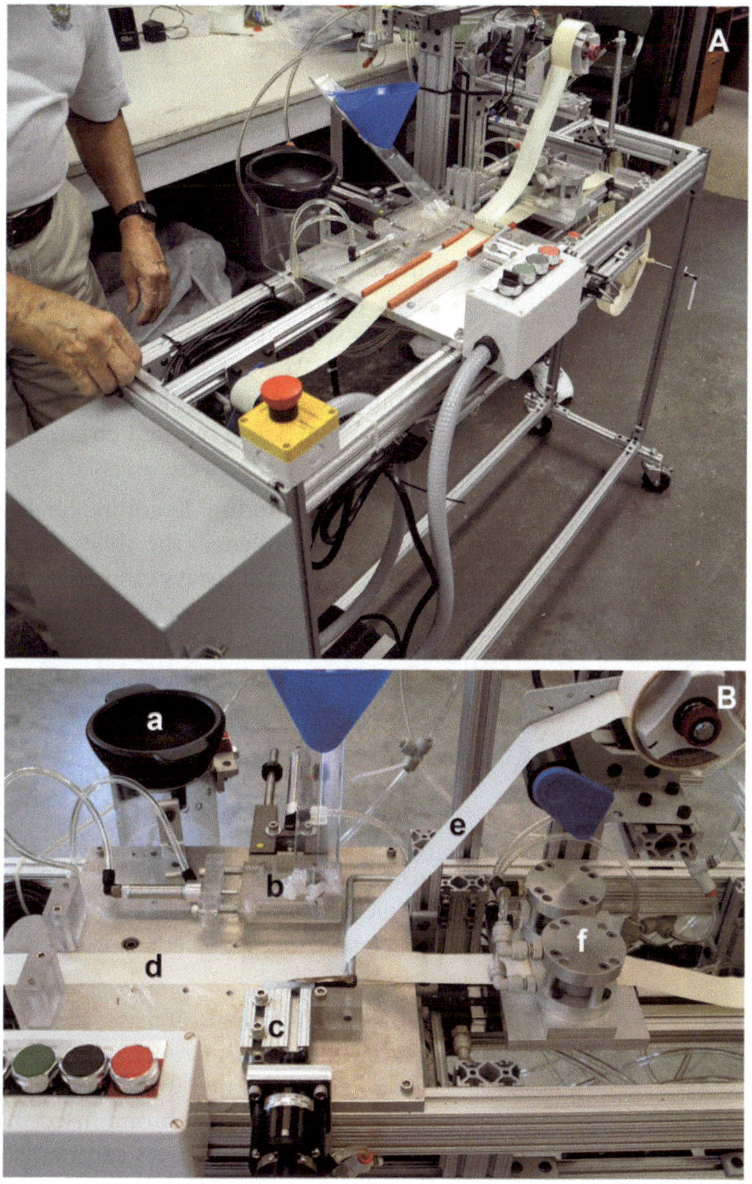


**Fig. 8** Cadaver packing machine showing two masking tape sources (**A**). The components of the machine (**B**) packs infected cadavers of *Tenebrio molitor* larvae by moving them from the cadaver holder (*a*) to a mechanical positioner (*b*), which positions the cadavers on a masking tape stripe (*d*). A second masking tape stripe (*e*) is rolled over the cadavers by a mechanical arm (*c*). A press (*f*) sets the two masking tape stripes glued in between cadavers while at the same time moves the stripe to set the next cadaver

cadavers in two strips; the two lines of tape are set together by a press which compresses the tape in between cadavers. The machine can operate continuously as long as it has enough tape and cadavers. Packed cadavers can be rolled and cut in sections for storage or delivery.

## 4 Notes

1. It is important to ensure a high level of quality control when producing EPNs for experimental or commercial purposes. One measure of quality control is percentage of viable IJs. A sample of IJs in aqueous suspension is placed onto Petri dish and the number of live or dead IJs is determined by prodding each nematode with a dissecting needle. Note that IJs may appear straight when they are stressed but will respond when prodded. At least 50 IJs should be counted and the count should be repeated at least three times. It may be desirable to create a lined grid on the Petri dish to assist in keeping track of which nematodes were assessed.
2. Another method to determine quality and fitness of EPNs is by measuring virulence (the ability to cause disease/mortality in a target host). The assays can follow similar procedures described earlier for inoculating insect hosts in Petri dishes on filter paper (*see* Subheading 3.1). Yet, assays conducted in soil or sand will give a more real-world assessment. Most EPNs should be able to kill a high percentage of *G. mellonella* (>90%) if using 50–200 IJs per insect; if this is not achieved then the virulence of the EPNs may be compromised due to various reasons such as environmental conditions, poor nutrition, strain deterioration (see below), etc. Given that *G. mellonella* is exceptionally susceptible to EPNs, a less susceptible host would provide a more precise virulence assessment (such as *T. molitor*, for which 200–800 IJs per insect may be needed or a specific commercial target pest can be used).
3. EPN quality can also be assessed by measuring the number of IJs invading a host (*see* ref. 6). A known quantity of IJs is placed on a Petri dish lined with filter paper, e.g., 100 IJs in 0.35 ml on a 60 cm dish, with one insect host such as *G. mellonella*. After 1–3 days the insect can be dissected and the number of invading nematode relative to the total applied is counted. In general, for most EPN species, one can expect 3–5% invasion for heterorhabditids and 5–10% for steinernematids. To facilitate the dissection, the insects can be digested using a pepsin solution first (*see* ref. 6).
4. A high level of contaminants in the harvested IJ suspension (e.g., fungi, protists, etc.) may be detrimental to nematode longevity or quality. Therefore, a decontamination step may be beneficial in such cases, and some producers may do this routinely. Previously many laboratories used 0.1% formalin, and some may still use it, but generally this has fallen out of favor due to safety concerns. An alternative is to use a low concentra-

tion of sodium hypochlorite (e.g., 1–3 %). In extreme cases, the nematodes may need to be surface sterilized. This can be accomplished by placing IJs in 5–10 % sodium hypochlorite or 0.1 % hyamine (methylbenzethonium chloride) for 15–20 min and then washing 1–3 times.

5. Quality and longevity of harvested nematodes can also be compromised by the presence of dead IJs or other nematode stages in the suspension; ideally suspensions should be maintained at ≥95 % live IJs. Dead IJs and other nematode stages can be separated by screening them out. A fine mesh screen (e.g., 500 mesh) can be partially immersed in a vessel with bubbling water, and the mixed suspension is applied on top of the screen. The fine mesh will allow IJs to pass through while dead IJs and larger nematode stages remain on the screen. Some fine cloths or tissue papers will also work for this purpose.
6. If IJs are tending to stick together in suspension a surfactant might be added such as a drop of Triton X-100; if the nematodes are clumping due to fungal contamination, sodium bicarbonate (1 g per 50 ml) may be added (*see* ref. 6).
7. Another issue that can impact nematode quality and fitness is beneficial trait deterioration (also called strain deterioration or attenuation) (*see* refs. 1, 19). Trait deterioration results from repeated culturing and is due to inbreeding depression or other genetic or nongenetic causes. Therefore, as much as possible, it is imperative to minimize the number of passages through any given host. Additionally, trait deterioration can be deterred through the creation of homozygous inbred lines (*see* refs. 1, 19).

## References

1. Shapiro-Ilan DI, Han R, Qiu X (2014) Production of entomopathogenic nematodes. In: Morales-Ramos J, Rojas G, Shapiro-Ilan D (eds) Mass production of beneficial organisms: invertebrates and entomopathogens. Academic, Amsterdam, pp 321–356
2. Dillman AR, Chaston JM, Adams BJ, Ciche TA, Goodrich-Blair H, Stock SP, Sternberg PW (2012) An entomopathogenic nematode by any other name. PLoS Pathog 8, e1002527
3. Lewis EE, Clarke DJ (2012) Nematode parasites and entomopathogens. In: Vega FE, Kaya HK (eds) Insect pathology, 2nd edn. Elsevier, Amsterdam, pp 395–424
4. Shapiro-Ilan DI, Gaugler R (2002) Production technology for entomopathogenic nematodes and their bacterial symbionts. J Ind Microbiol Biotech 28:137–146
5. White GF (1927) A method for obtaining infective nematode larvae from cultures. Science 66:302–303
6. Kaya HK, Stock SP (1997) Techniques in insect nematology. In: Lacey LA (ed) Manual of techniques in insect pathology. Academic, San Diego, CA, pp 281–324
7. Shapiro-Ilan DI, Gaugler R, Tedders WL, Brown I, Lewis EE (2002) Optimization of inoculation for *in vivo* production of entomopathogenic nematodes. J Nematol 34:343–350
8. Morales-Ramos JA, Rojas MG, Kay S, Shapiro-Ilan DI, Tedders WL (2012) Impact of adult weight, density, and age on reproduction of *Tenebrio molitor* (Coleoptera: Tenebrionidae). J Entomol Sci 47:208–220
9. Morales-Ramos JA, Rojas MG, Shapiro-Ilan DI, Tedders WL (2013) Use of nutrient self-selection as a diet refining tool in *Tenebrio moli-*

*tor* (Coleoptera: Tenebrionidae). J Entomol Sci 48:206–221

10. Morales-Ramos JA, Rojas MG, Shapiro-Ilan DI, Tedders WL (2010) Developmental plasticity in *Tenebrio molitor* (Coleoptera: Tenebrionidae): analysis of instar variation in number and development time under different diets. J Entomol Sci 45:75–90
11. Morales-Ramos JA, Rojas MG, Shapiro-Ilan DI, Tedders WL (2011) Self-selection of two diet components by *Tenebrio molitor* (Coleoptera: Tenebrionidae) larvae and its impact on fitness. Environ Entomol 40:1285–1294
12. Morales-Ramos JA, Rojas MG, Shapiro-Ilan DI, Tedders WL (2011) Automated insect separation system. US Patent 8,025,027 B1
13. Shapiro-Ilan DI, Rojas MG, Morales-Ramos JA, Tedders WL (2012) Optimization of a host diet for in vivo production of entomopathogenic nematodes. J Nematol 44:264–273
14. Gaugler R, Brown I, Shapiro-Ilan DI, Atwa A (2002) Automated technology for i*n vivo* mass production of entomopathogenic nematodes. Biol Control 24:199–206
15. Shapiro-Ilan DI, Tedders WL, Morales-Ramos JA, Rojas MJ (2014) System and method for producing beneficial parasites. US Patent 8,677,935
16. Shapiro-Ilan DI, Lewis EE, Behle RW, McGuire MR (2001) Formulation of entomopathogenic nematode-infected-cadavers. J Invertebr Pathol 78:17–23
17. Shapiro-Ilan DI, Tedders WL, Lewis EE (2008) Application of entomopathogenic nematode-infected cadavers from hard-bodied arthropods for insect suppression. US Patent 7,374,773
18. Morales-Ramos JA, Tedders WL, Dean CB, Shapiro-Ilan DI, Rojas MG (2013) Apparatus for packing arthropods infected with entomopathogenic nematodes. US Patent 8,505,236 B1
19. Bai C, Shapiro-Ilan DI, Gaugler R, Hopper KR (2005) Stabilization of beneficial traits in *Heterorhabditis bacteriophora* through creation of inbred lines. Biol Control 32:20–227

# Part VII

## Monitoring of Applied Microbes

# Chapter 12

# Detection and Quantification of the Entomopathogenic Fungal Endophyte *Beauveria bassiana* in Plants by Nested and Quantitative PCR

**Inmaculada Garrido-Jurado, Blanca B. Landa, and Enrique Quesada-Moraga**

## Abstract

The described protocol allows detecting as low as 10 fg the entomopathogenic fungal endophyte *Beauveria bassiana* in host plants by using a two-step nested PCR with the ITS1F/ITS4 and BB.fw and BB.rv primer pairs. On the other hand, a qPCR protocol using BB.fw and BB.rv primers is also available allowing the quantification of up to 26 fg of *B. bassiana* DNA per 20 ng of leaf DNA.

**Key words** Mitosporic ascomycetes, DNA, Leaves, Nested PCR, Quantitative PCR

## 1 Introduction

The entomopathogenic mitosporic ascomycete (EMA) *Beauveria bassiana* have been reported to be naturally a fungal endophyte [1], and also it has been artificially applied to plants conferring systemic protection against insect pests. First, it was supposed to act through antibiosis or feeding deterrence [2]. However, the latest researches shed new insight on their colonization on plant tissues and transmission [3], and consequently on the resistant that provided to the plant against pests and fungal pathogens [4, 5]. *In-planta* detection of this EMA becomes an important issue. Therefore, the present chapter develops the methodology to detect and quantify the EMA *Beauveria bassiana* in plants. For that, the combination of the universal ITS1F/ITS4 and the *B. bassiana* specific BB.fw and BB.rv primers is used in a two-step nested PCR, allowing the amplification of up to 10 fg of *B. bassiana*. The detection protocol developed on artificially inoculated opium poppy *Papaver somniferum* L. (Papaveraceae) as plant host has been also tested on very different hosts such as Canary Island date palm *Phoenix canariensis* Chab. (Palmaceae). On the other hand, a qPCR protocol using BB.fw and

Travis R. Glare and Maria E. Moran-Diez (eds.), *Microbial-Based Biopesticides: Methods and Protocols*, Methods in Molecular Biology, vol. 1477, DOI 10.1007/978-1-4939-6367-6_12, 

BB.rv primers is able to quantify up to 26 fg of *B. bassiana* DNA per 20 ng of leaf DNA. PCR and qPCR can be used to study entomopathogenic endophytic fungus–plant interactions.

## 2 Materials

### 2.1 Disinfection of Plant Material

1. 70% ethanol.
2. 2% aqueous solution of sodium hypochlorite.
3. Selective culture medium: 20.0 g of Sabouraud Glucose Agar (SDA) supplemented with 500 mg/l chloramphenicol, 500 mg/l streptomycin sulfate, 500 mg/l ampicillin, and 500 mg/l dodine 65 WP. Make up to 1 l with water.
4. Sterilized water.
5. Sterile filter paper.
6. Sterile scalpel.

### 2.2 Components for DNA Plant Extraction

1. Lyophilized leaves.
2. Glass beads (0.425–0.600 mm).
3. 2-ml screw-cap tubes.
4. FastPrep beater (Bio 101 Systems, Qbiogene) or similar.
5. Extraction buffer: 100 mM Tris–HCl pH 8, 50 mM EDTA in sterile distilled $H_2O$, 500 mM NaCl, 10 mM mercaptoethanol, and 1% (w/v) SDS.
6. RNase A solution (5 μg/μl).
7. Phenol/chloroform/isoamyl alcohol (25:24:1).
8. 3 M NaCl.
9. Absolute isopropanol.
10. 80% ethanol.
11. Elution buffer: 10 mM Tris–HCl pH 8, 1 mM EDTA in sterile distilled $H_2O$.
12. Nanodrop Spectrophotometer ND2000 (Thermo Scientific).

### 2.3 Two-Steps Nested-PCR Amplification

1. Template DNA.
2. 10× Reaction buffer B without $MgCl_2$ (Solis BioDyne).
3. 25 mM $MgCl_2$ (Solis BioDyne).
4. dNTP mix 20 mM of each (Solis BioDyne).
5. Primer pairs that belong to ITS1-5.8S-ITS2 regions of the rDNA (Fig. 1, Table 1).
6. DNA Polymerase (Solis BioDyne).
7. Purified *Beauveria bassiana* genomic DNA.
8. Ultrapure sterile water.

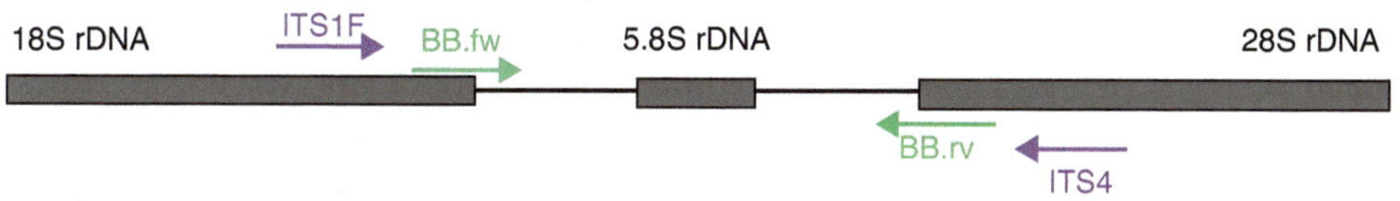


**Fig. 1** Relative positions of the four primers in the rDNA. Primers in *purple* and *green* are used in the first and second reaction of the two-step nested PCR, relatively

**Table 1**
**Primers used to detect endophyte Beauveria bassiana in plants**

| Two-step nested-PCR | Primers | | References |
|---|---|---|---|
| | Names | Sequences | |
| First step | ITS4 | TCCTCCGCTTATTGATATGC | [6] |
| | ITS1F | CTTGGTCATTTAGAGGAAGTAA | [7] |
| Second step | BB.fw | GAACCTACCTATCGTTGCTTC | [8] |
| | BB.rv | ATTCGAGGTCAACGTTCAG | [8] |

#### 2.4 Agarose Gel Components

1. Agarose LOW EEO.
2. 10× Tris Acetate EDTA (TAE) buffer, pH 8.0: 48.40 g Tris, 11.42 ml Glacial acetic acid, 3.72 g EDTA (disodium salt), pH 8 in 1 l Distilled water.
3. SYBR Safe DNA gel stain (10,000×).
4. Combs and a horizontal electrophoresis system (Bio-Rad) or similar.
5. Loading buffer: 30% glycerol (w/v), 0.1 M EDTA pH 8, 1.0% SDS, 0.25% bromophenol blue (w/v), 0.25% cyanol xylene (w/v).
6. 100-bp molecular weight ladder (Solis BioDyne) or similar.

#### 2.5 Real-Time PCR Quantification

1. DNA sample.
2. iQ SYBR Green Supermix (BioRad) or similar.
3. iCycler IQ apparatus (Bio-Rad) or similar.

## 3 Methods

#### 3.1 Disinfection of Plant Material

1. Disinfect externally the leaves with 70% ethanol (v/v) for 2 min, 1% sodium hypochlorite for 5 min, and rinse twice in sterile distilled water (*see* **Note 1**).
2. Dry on sterile filter paper under sterile airflow.
3. Cut 2 cm$^2$ fragments with a sterile scalpel and freeze immediately (*see* **Notes 2** and **3**).

### 3.2 DNA Plant Extraction

1. Grind 5 mg of lyophilized leaf with the same weight of glass beads (0.425–0.600 mm) into 2 ml screw-cap tubes in a FastPrep beater for 30 s at a run speed of 5.5 m/s.
2. Add 490 μl of extraction buffer and 5 μl of RNase A solution into sample tube, mix by vigorous shaking, and incubate at 65 °C for 30 min. Mix three or four times during incubation by inverting the tube.
3. Add an equal volume of phenol/chloroform/isoamyl alcohol (25:24:1) in one step to the lysate and mix gently.
4. Centrifuge at room temperature (RT), 9402 × *g* for 10 min.
5. Transfer the aqueous top layer into a new 2 ml-tube (*see* **Note 4**).
6. Add 650 μl of 2 mol/l NaCl, incubate at RT for 30 min, and add an equal volume of absolute ice-cold isopropanol.
7. Centrifuge at RT/9300 × *g* for 12 min.
8. Discard the supernatant and add 200 μl of 80% ethanol.
9. Centrifuge at RT/13,400 × *g* for 2 min.
10. Decant the tube and dry the pellet inside the burn-in chamber under sterile airflow for 30 min.
11. Dissolve in 50 μl of elution buffer or ultrapure sterile water.
12. Test concentration and quality of DNA using a Nanodrop Spectrophotometer. An A260/A280 ratio of 2.0 ± 0.1 indicates a clean preparation of DNA.

### 3.3 Two-Step Nested-PCR Amplification

1. Perform the first reaction of nested PCR amplification with primer pairs ITSF1/ITS4, using 1 μl of template DNA (20 ng/μl) in a 25-μl reaction volume. Add 2.5 μl of 10× reaction buffer, 1.5 mM $MgCl_2$, 50 μM of each dNTP, 1.0 μM of each primer, 1.5 units of DNA Polymerase.
2. Use purified *B. bassiana* DNA (1 ng) and ultrapure sterile water as positive and negative controls, respectively.
3. Set the cycling program as an initial denaturation step of 4 min at 95 °C, followed by 35 cycles of 1 min denaturation at 95 °C, 1 min annealing at 61 °C, and 1 min extension at 72 °C and a final 10 min extension step at 72 °C followed by a 4 °C soak.
4. Use 1 μl of this PCR product for the second round of amplification using primer pairs BB.fw/BB.rv, at the same amounts in a 25-μl reaction volume and the same optimized amplification conditions as in Subheading 3.3, **step 1**.

### 3.4 Agarose Gel Electrophoresis

1. Weight 1.5 g of agarose, add 100 ml of 1× TAE, and heat until the mix is melted.
2. After cooling, add 5 μl of SYBR Safe DNA gel stain (*see* **Note 5**).

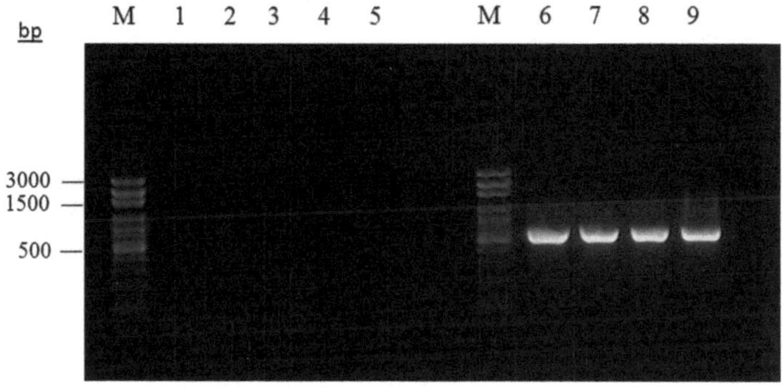


**Fig. 2** Detection of *Beauveria bassiana* in artificially infected plants using two-step nested. *M*, 100 bp molecular weight ladder; *lane 1*, no DNA template; *lanes 2* and *3*, DNA extract for noninoculated opium poppy leaves; *lanes 4* and *5*, DNA extract for noninoculated palm leaves; *lane 6*, DNA extract from surface-disinfected opium poppy leaves inoculated with EABb 04/01-Tip *B. bassiana*; *lane 7*, DNA extract from surface-disinfected palm leaves inoculated with EABb 04/01-Tip *B. bassiana*; *lane 8*, DNA extract from surface-disinfected palm leaves inoculated with EABb 12/01-Rf *B. bassiana*; *lane 9*, DNA extract from surface-disinfected palm leaves inoculated with EABb 07/06-Rf *B. bassiana*

3. When the gel is polymerized, remove the combs and place the gel in a horizontal electrophoresis system containing enough 1× TAE to cover agarose gel.
4. Mix 5 μl each sample with 1 μl of loading buffer. The band size is estimated by comparison with a 100 bp molecular weight ladder disposed in the first slot.
5. Amplification products are separated by electrophoresis for 60 min at 80 V.
6. Gel is observed under UV light (260–302 nm) (*see* Fig. 2).

### 3.5 Real-Time PCR Quantification

1. Prepare the qPCR reaction mixture (final volume of 20 μl) containing: 1 μl of DNA sample, 2× iQ SYBR Green Supermix (BioRad), and 0.3 μM of each BB.fw/BB.rv primer.
2. Each sample should be represented by two or three technical replicates.
3. Run plate in iCycler IQ apparatus using the following PCR program: initial denaturation at 95 °C for 2 min followed by 35 cycles of 2 min at 95 °C, 1 min at 65 °C, 1 min at 72 °C, and 83 °C for 15 s followed by a final extension step at 72 °C for 10 min to end. Optional: a melting curve analysis to test for specificity of amplification and absence of primer dimers can be performed by heating to 95 °C, cooling to 72 °C, and slowly heating to 95 °C at 0.5 °C every 10 s, with continuous measurement of fluorescence at 520 nm.

## 4 Notes

1. Increase the percentage of sodium hypochlorite (up to 3.25 %) and the exposition time (up to 10 min) to disinfected leaves with a strong sclerotization. In particularly difficult cases use a three-step ethanol, sodium hypochlorite, ethanol treatment (*see* ref. 9).
2. The 2 $cm^2$ fragment should preferably be taken from the leaf vascular tissue close to the petiole.
3. One hundred microliter of the last rinse of each sample should be cultured on petri plates containing selective culture medium to ensure total disinfestations of the leaves.
4. Use wire bore tips; cutting normal ones and autoclave. This avoids disturbing the interphase.
5. Cover the electrophoresis system containing the stained gel with an aluminum foil, since SYBR safe is photosensitive.

## Acknowledgments

This research was supported by project AGR-7681 from the Consejería de Innovación Ciencia y Empresa de la Junta de Andalucía, Spain.

### References

1. Herrero N, Dueñas E, Quesada-Moraga E, Zabalgogeazcoa I (2012) Prevalence and diversity of viruses in the entomopathogenic fungus *Beauveria bassiana*. Appl Environ Microbiol 78:8523–8530
2. Vega FE (2008) Insect pathology and fungal endophytes. J Invertebr Pathol 98:277–279
3. Quesada-Moraga E, López-Díaz C, Landa BB (2014) The hidden habit of the entomopathogenic fungus *Beauveria bassiana*: first demonstration of vertical plant transmission. PLoS One 9:e89278
4. Vega FE, Goettel MS, Blackwell M, Chandler D, Jackson MA, Keller S, Koike M, Maniania NK, Monzón A, Ownley BH, Pell JK, Rangel DEN, Roy HE (2009) Fungal entomopathogens: new insights on their ecology. Fungal Ecol 2:149–159
5. Ownley BH, Gwinn KD, Vega FE (2010) Endophytic fungal entomopathogens with activity against plant pathogens: ecology and evolution. BioControl 55:113–128
6. White TJ, Bruns T, Lee S, Taylor JW (1990) Amplification and direct sequencing of fungal ribosomal RNA genes for phylogenetics. In: Innis MA, Gelfand DH, Sninsky JJ, White TT (eds) PCR protocols: a guide to methods and applications. Academic, New York, pp 315–322
7. Gardes M, Bruns TD (1993) ITS primers with enhanced specificity for basidiomycetes—application to the identification of mycorrhizae and rusts. Mol Ecol 2:113–118
8. Landa BB, Lopez-Diaz C, Jimenez-Fernandez D, Montes-Borrego M, Munoz-Ledesma FJ, Ortiz-Urquiza A, Quesada-Moraga E (2013) *In-planta* detection and monitorization of endophytic colonization by a *Beauveria bassiana* strain using a new-developed nested and quantitative PCR-based assay and confocal laser scanning microscopy. J Invertebr Pathol 114:128–138
9. Bills GF (1996) Isolation and analysis of endophytic fungal communities from woody plants. In: Redlin SC, Carris LM (eds) Endophytic fungi in grasses and woody plants: systematics, ecology, and evolution. American Phytopathological Society (APS Press), Minnesota, pp 31–65

# Chapter 13

# Plant Tissue Preparation for the Detection of an Endophytic Fungus *In Planta*

**Aimee C. McKinnon**


## Abstract

The identification of fungal endophytes often relies on culturing isolates from surface-sterilized plant tissue. However, molecular techniques have enabled the rapid detection and identification of targeted endophyte species, and next-generation sequencing technology provides an opportunity to obtain comprehensive information on endophytic communities, directly from plant tissue. In order to achieve accurate detection from internal tissues, surface microbes and associated deoxyribonucleic acid (DNA) must be eliminated, with particular consideration for the type of plant tissue and the efficacy of the surface sterilization procedure used. The methodology described later was developed specifically for detection of DNA from the entomopathogenic fungal endophyte *Beauveria bassiana* (Vuillemin) (Ascomycota: Hypocreales) in various tissues of *Zea mays* (L.). However, the protocol may be easily applied to other fungi and bacterial endophytes. Included is a stringent sodium hypochlorite-based surface sterilization protocol for plant material in preparation for polymerase chain reaction (PCR) to detect target DNA within plant tissue. Included are a treatment for dealing with surface DNA contamination and a novel procedure for assessing the efficacy of surface sterilization using PCR.


**Key words** Entomopathogenic fungal endophyte, Surface sterilization, Biocontrol, PCR, Next-generation sequencing, DNA isolation, Propidium monoazide

## 1 Introduction

There are a number of challenges presented for effective isolation or recovery of endophytic fungi from living plants. For example, the conidia produced from genera of Ascomycota may withstand more duress than plant tissue, especially compared with soft foliage and fine roots. Yet, surface sterilization methodology employed for endophyte isolation is often selected to maintain the structural integrity of the plant tissue, at the risk of not eliminating all of the viable epiphytic propagules [1]. This may confound results, particularly in experiments which seek to recover endophytic strains from inoculated plants. Resident epiphytes found on host surfaces may also live in close association with the plant, penetrating superficial epidermal layers without becoming truly endophytic per se. Indeed,

Travis R. Glare and Maria E. Moran-Diez (eds.), *Microbial-Based Biopesticides: Methods and Protocols*, Methods in Molecular Biology, vol. 1477, DOI 10.1007/978-1-4939-6367-6_13, © Springer Science+Business Media New York 2016

the distinction between endophytic and epiphytic fungi may be difficult to determine even with modern microscope technology [2]. For biocontrol purposes, the inundative application of the fungal endophyte to external plant surfaces increases the likelihood of surface contamination after sterilization [3]. Additionally, hydrophobic surfaces (plant, fungal) may be protected by micro-sized air bubbles during submersion in a disinfectant [4]. As it only takes one viable spore to germinate and yield a false 'endophyte' positive, surface sterilization efficacy should be carefully considered and tested within each unique system, in order to claim whether the inoculum applied has become endophytic [5].

With these challenges in mind, the surface sterilization method employed requires a suitable experimental control to confirm efficacy. Often, the control for surface sterilization has involved pipetting 100 μL of the final wash water and plating it on selective media [6]. Though this may be an effective control with some fungal (or bacterial) species, for *Beauveria* spp., this control is insufficient because of the highly hydrophobic nature of the infective conidia [7] and/or because hyphal structures may weave tightly through epidermal layers allowing the fungus to bind securely to the plant surface [8]. Consequently, if even a few epiphytic *Beauveria* propagules survive the sterilization procedure, they are more likely to be adhering to the plant surface and not floating freely in the wash water. Since the advent of molecular methods for endophyte detection, more challenges arise for plant tissue preparation. In particular, surface sterilization efficacy must be complete to eliminate both viable epiphytes and nonviable propagules adhering to plant surfaces. Furthermore, it is established that DNA from dead organisms can still be amplified by PCR [9]. Therefore, the determination of endophytic status cannot simply rely on surface sterilization to kill surface microorganisms. Propidium monoazide (PMA™, Biotium) is a photoreactive dye that is able to intercalate the DNA in ruptured (nonviable) cells rendering it unavailable for PCR amplification. After surface sterilization, PMA can be used to treat plant material prior to tissue grinding and DNA isolation. The advantage of using PMA is that the molecule possesses an azide group, which is activated readily with light exposure, enabling covalent cross-linkage with available organic moieties such as nucleic acid [10]. The reaction is thus fixed with sufficient light exposure, and subsequent DNA from intact cells (such as from plant and symbiotic microorganisms) remains free for PCR amplification. The inclusion of a PMA binding step for endophyte detection therefore requires stringent surface sterilization to ensure all surface inoculum are damaged, as DNA from nonviable intact conidia/cells on the surface will result in amplification. Furthermore, the inclusion of a PMA treatment prior to extraction may enhance the sensitivity of the PCR, because DNA from damaged plant cells is also affected by the dye and thus the ratio of plant to target (endophyte) DNA following treatment is decreased.

Described below is a protocol optimized for plant tissue preparation prior to DNA extraction to enhance PCR detection of artificially inoculated *Beauveria* sp. endophytes *in planta*.

## 2 Materials

The following surface sterilization method requires the sodium hypochlorite solution to be prepared fresh on the day of processing. All materials used for this protocol should be sterile/autoclaved as appropriate and handled aseptically to avoid introducing contamination.

1. To prepare 2.5 % sodium hypochlorite bleach solution (per 250 mL), mix: 50 mL 0.05 % Tween 80, 75 mL sterile distilled water, 125 mL 5 % active available chlorine sodium hypochlorite.
2. Hydrogen peroxide wet wipes (e.g., Clorox Professional hydrogen peroxide 3 % wet wipes).
3. Permeable cloth bags or metal tea strainers (*see* **Note 1**).
4. 500 mL volume sterile plastic containers with lids.
5. *Beauveria* semiselective media (BSM): quarter strength potato dextrose agar, 350 mg/L streptomycin sulfate, 50 mg/L tetracycline hydrochloride, and 125 mg/L cyclohexamide (*see* ref. 11).
6. *Beauveria* sp. or other fungal conidial suspension.
7. Propidium monoazide (PMA™, Biotium) 20 mM stock solution (prepared and stored according to the manufacturer's instructions).
8. Onion (*Allium* sp.) epidermis, peeled manually from layers and cut into 12 × 1 cm$^2$ pieces.
9. 1 % agar (solid media in standard petri dishes, 6× plates).
10. Halogen light lamp, ≥600 W.

## 3 Methods

### *3.1 Assessing Surface Sterilization Efficacy*

The onion epidermis assay was developed in order to adequately assess the efficacy of surface sterilization for PCR detection. Onion epidermis can be manually peeled to a single cell layer, providing a plant surface which enables 'epiphytic' adhesion of the inoculum while excluding the possibility of endophytic colonization or passive absorption of the inoculum by the plant (which may occur with live organs such as roots when using a conidial suspension). Additionally, onion epidermis is ideal for microscopic visualization when optimization of the protocol is required. For example, the inoculum can be applied in suspension to the onion epidermis 20, 72, or 96 h prior to DNA isolation and be subsequently visualized before and after surface sterilization directly on the epidermal surface. The number of intact conidia/hyphal bodies remaining after sterilization (viable and/or nonviable) can be estimated per centimeter squared area of epidermis, and PCR/qPCR experiments can be used to determine the detection threshold for conidia per $cm^2$, thereby reducing the likelihood of detecting confounding surface inoculum in subsequent experiments. To assess efficacy:

1. Prepare an inoculum suspension of the target endophyte. The same load of inoculum can be applied as used in corresponding experiments or from *ca.* $10^5$ to $10^7$ conidia/mL will provide an appropriate quantity.
2. Peel single epidermal layers from an onion and cut into twelve 1 $cm^2$ pieces.
3. Place the pieces on 1 % agar, two per plate.
4. Inoculate the pieces with 25 μL of suspension.
5. Incubate in the dark for 20 h at ambient temperature.
6. Surface sterilize the epidermis pieces using the same procedure selected for the endophyte detection experiments (*see* Subheading 3.2) (*see* **Note 2**).
7. Following surface sterilization, half of the epidermis samples can be treated with propidium monoazide (PMA) to demonstrate the efficacy of the dye for concordant PCR experiments (*see* Subheading 3.3) (*see* **Note 3**).
8. To ascertain inoculum viability after surface sterilization, extra inoculated epidermis can be included, processed according to the protocol described in Table 1 and incubated directly on growth media prepared as appropriate for the inoculum.

**Table 1**
**Surface sterilization methods for inoculated plant tissue**

| Sample | 85 % EtOH duration (min) | $sdH_2O$ rinse[a] (s) | 2.5 % NaOCl duration[b] (min) | $sdH_2O$ washes[b] |
|---|---|---|---|---|
| Leaves, stems, roots, epidermis | 1 | 30 | 5 | 1 min × 2 |
| Seed | 1 | 30 | 7 | 1 min × 2 |

[a]A rinse in water is recommended between the EtOH and NaOCl incubation steps to protect plant tissue from extended exposure in disinfectant

[b]Duration in bleach may vary depending on the plant species or age of plant (*see* **Note 4**)

### 3.2 Surface Sterilization of Inoculated Plant Tissue

1. Take desired samples (leaves, stems, or roots) and gently remove any adhering soil or dust (i.e., in tap water).
2. Gently wipe leaf, stem, or root surfaces with hydrogen peroxide wipes (*see* **Note 5**).
3. Cut samples into 3 cm length fragments.
4. Put leaves, stems, and root samples into permeable bags or tea strainers.
5. Wash the bags/strainers containing samples gently in 0.05% Tween 80, this is important to break the surface tension, penetrate the surface, and disperse air bubbles. Drain carefully.
6. Transfer the bags/strainers into a sterile area such as a laminar flow cabinet and start the surface sterilization procedure (*see* Table 1). Ensure complete submersion of plant tissue during incubation.
7. Transfer sterilized tissue to a sterile petri plate.
8. Cut samples further with a sterile blade, to 1 cm long fragments (or 1 $cm^2$ for leaves), if necessary.
9. Transfer individual samples each into 500 μL $sdH_2O$ in clear 0.6 mL centrifuge tubes.
10. Store samples at 4 °C as necessary (*see* **Note 6**) and/or proceed to the PMA treatment protocol (*see* Subheading 3.3).

### 3.3 Protocol for Treating Plant Samples with PMA for PCR

The following protocol was adapted by Wisnu Wicaksono (pers. comms) from the manufacturer's instructions (Biotium, PMA™ dye Product Information sheet). Refer also to Nocker et al. (*see* refs. 9, 10, 12) for PMA treatment methodology.

1. To the surface-sterilized plant samples (*see* **Note 7**) in clear 0.6 mL centrifuge tubes, pipette 1.25 μL PMA per 500 μL (the volume prepared per tube).
2. Vortex the tubes briefly, ensure full submersion of the samples in the solution by flicking the tube and transfer immediately to a container/sealed box in order to incubate in the dark for 5 min (*see* **Note 8**).
3. Transfer the tubes to a foil-lined tray on ice and expose to halogen light lamp for 5 min: The lamp should be set 20 cm directly from the samples and the samples turned at least once during the 5 min exposure. Alternatively, a shaking table may also be utilized to provide continuous mixing of the solution during light exposure as per the product information sheet.
4. Store at 4 °C in solution. Prior to grinding the plant sample with liquid nitrogen for the isolation of DNA, decant the solution using a pipetman.

## 4 Notes

1. Permeable cloth bags can be prepared using Miracloth or similar. For the tea strainers, the stainless steel infuser tongs with the wire mesh infuser and handles are ideal as they do not float or react with the bleach and can be handled aseptically.
2. Occasionally, onion epidermis may degrade with the described protocol. In that instance, more treated epidermis can be processed than required to ensure a sufficient number of treated samples.
3. Epidermis pieces should be treated individually with PMA dye to ensure a complete reaction with the plant surface. However, the samples can be pooled together to two epidermis pieces per DNA extraction to maximize yield.
4. For example, if a 5 min interval in 2.5 % bleach degrades soft tissues, break the incubation time to 2 × 3 min intervals with a rinse step in between. Plants samples that have been heavily inoculated prior may require more time and/or a higher bleach concentration (for example, stems and roots may require 7 min in 3.25 % F.A.C. NaOCl).
5. Certain plant tissues, such as roots with root hairs and/or other fine structures may break with the mechanical disturbance from wiping. In this instance, **step 2** in Subheading 3.2 (with hydrogen peroxide wipes) can be excluded from the procedure.
6. Newly surface sterilized samples can be stored temporarily (e.g., up to 10 days) but it is recommended to proceed to treatment and DNA extraction immediately.
7. To demonstrate the efficacy of the PMA treatment, it is recommended to include additional experimental controls: Heat treat four 500 μL aliquots of a conidial suspension to 85 °C for 10 min. Additionally, include a further four suspensions containing hyphal material and treat these using the selected surface sterilization protocol as the plant material (*see* ref. 13). Half of the heat-treated and sterilized samples can then be processed with PMA. Finally, these eight suspensions can be included as DNA positive samples for comparison in subsequent PCR experiments to confirm that the amplification of DNA from sterilized fungal matter is completely hindered by the PMA dye.
8. To limit light exposure during the reaction phase (dark incubation), it is recommended to process no more than ten samples for PMA treatment at a time.

## References

1. Hyde KD, Soytong K (2008) The fungal endophyte dilemma. Fungal Divers 33:163–173
2. Wilson D (1995) Endophyte—the evolution of a term, and classification of its use and definition. Oikos 73(2):274–276
3. Eilenberg J, Hajek A, Lomer C (2001) Suggestions for unifying the terminology in biological control. BioControl 46(4):387–400
4. Schulz B, Boyle C (2005) The endophytic continuum. Mycol Res 109:661–686
5. Schulz B, Krohn K, Meier K, Draeger S (2011) Isolation of endophytic fungi for the production of biologically active secondary metabolites. In: Pirttilä AM, Sorvari S (eds) Prospects and applications for plant-associated microbes. A laboratory manual, Part B: Fungi, 1st edn. Biobien Innovations, Germany
6. Hallmann J, Berg G, Schulz B (2006) Microbial root endophytes. In: Czeschlik D (ed) Soil biology. Springer, Berlin
7. Zhang SZ, Xia YX, Kim B, Keyhani NO (2011) Two hydrophobins are involved in fungal spore coat rodlet layer assembly and each play distinct roles in surface interactions, development and pathogenesis in the entomopathogenic fungus, *Beauveria bassiana*. Mol Microbiol 80(3):811–826
8. Wagner BL, Lewis LC (2000) Colonization of corn, Zea mays, by the entomopathogenic fungus *Beauveria bassiana*. Appl Environ Microbiol 66(8):3468–3473
9. Nocker A, Cheung C-Y, Camper AK (2006) Comparison of propidium monoazide with ethidium monoazide for differentiation of live vs. dead bacteria by selective removal of DNA from dead cells. J Microbiol Methods 67(2):310–320
10. Nocker A, Sossa-Fernandez P, Burr MD, Camper AK (2007) Use of propidium monoazide for live/dead distinction in microbial ecology. Appl Environ Microbiol 73(16):5111–5117
11. Brownbridge M, Reay SD, Nelson TL, Glare TR (2012) Persistence of *Beauveria bassiana* (Ascomycota: Hypocreales) as an endophyte following inoculation of radiata pine seed and seedlings. Biol Control 61(3):194–200
12. Nocker A, Sossa KE, Camper AK (2007) Molecular monitoring of disinfection efficacy using propidium monoazide in combination with quantitative PCR. J Microbiol Methods 70(2):252–260
13. Petrini O (1986) Taxonomy of endophytic fungi of aerial plant tissues. In: van den Huevel J, Fokkema NJ (eds) Microbiology of the phyllosphere. Cambridge University Press, Cambridge, UK

# Part VIII

## Quality Control, Safety, and Registration

# Chapter 14

# Measuring Chitinase and Protease Activity in Cultures of Fungal Entomopathogens

Peter Cheong, Travis R. Glare, Michael Rostás, and Stephen R. Haines

## Abstract

Entomopathogenic fungi produce a variety of destructive enzymes and metabolites to overcome the unique defense mechanisms of insects. In a first step, fungal chitinases and proteinases need to break down the insect's cuticle. Both enzyme classes support the infection process by weakening the chitin barrier and by producing nutritional cleavage products for the fungus. In a second step, the pathogen can now mechanically penetrate the weakened cuticle and reach the insect's hemolymph where it starts proliferating. The critical enzymes chitinase and proteinase are also excreted into the supernatants of fungal cultures and can be used as indicators of virulence. Chromogenic assays adapted for 96-well microtiter plates that measure these enzymes provide a sensitive, fast, and easy screening method for evaluating the potential biocontrol activity of fungal isolates and may be considered as an alternative to laborious and time-consuming bioassays. Furthermore, monitoring fungal enzyme production in dependence of time, nutrient sources, or other factors can facilitate in establishing optimal growth and harvesting conditions for selected isolates with the aim of achieving maximum biocontrol activity.

**Key words** Proteases, Chitinase, Insect defense, Chromogenic assay, 96-Well plates, Sensitive, Fast, Easy

## 1 Introduction

### *1.1 Fungal Infection*

Unlike bacteria, protozoa, and viruses, entomopathogenic fungi do not rely on the insect to ingest the infective propagules and to penetrate via the midgut. Neither are specific entry points like oviposition tubes, spiracles, or wounds required [1, 2]. Infection through the external cuticle is the unique characteristic of true entomopathogenic fungi [2, 3]. The spores of a fungus will attach themselves onto an insect, germinate and after signal recognition, stop horizontal growth and initiate penetration [4]. Through turgor pressure and appressoria formation in some fungal species, the hyphae forcibly perforate the cuticle of the insect like needles when the surface barrier has already weakened and partially dissolved by enzymatic reactions and secreted organic acids (e.g. oxalic acid) [4–9]. The hyphae grow toward the nutrient-rich hemolymph of

Travis R. Glare and Maria E. Moran-Diez (eds.), *Microbial-Based Biopesticides: Methods and Protocols*, Methods in Molecular Biology, vol. 1477, DOI 10.1007/978-1-4939-6367-6_14, the insect and each invading hypha proliferates into yeast-like structures known as blastospores that disperse inside the hemolymph and evade the immune system of the insect [2, 10]. The blastospores will continue to grow into hyphae which then will form mycelia that eventually emerge from the cadaver of the insect.

### 1.2 Insect Defense

The cuticular surface of insects is covered by a thin layer of lipids, mainly hydrocarbons, wax esters, fatty alcohols, and free or esterified fatty acids [11], with crystalline microfibers of chitin embedded in a protein matrix which constitutes 55–80% of the insect cuticle structure [4, 12, 13]. The nonpolar structure, which is hydrophobic, plays a major role in preventing desiccation, altering the absorption of water, chemicals, and microorganism penetration. The structure also contains antifungal properties (toxic cuticular lipids, benzoquinones, small peptides, proteins, insect-derived chitinase, protease inhibitor, and antibiotics) which are produced by the insect or bacteria symbionts and secreted to the cuticle [14–20]. The molting fluid enzymes allow rapid ecdysis and molting, recycling of old cuticle, leaving the underlying layer of the new cuticle impervious to degradation [21], and the removal of any fungal material clinging onto the surface of the old cuticle [16]. The fluid also contains high level of protease inhibitor, at least two types in some [22], with high affinity for fungal proteases [21]. Other defenses against fungal pathogens include hydrogen peroxide self-medication via ingestion [23], glutathione-*S*-transferase (GST) and body fat to detoxify exogenous compounds [24], body thermoregulation [25], melanization [7], cannibalism, burial, avoidance, removal from social nesting environment voluntarily or as cadavers before sporulation [26] and within the hemolymph, the up-regulation of innate immune systems to engulf intruding pathogens [2]. All these mechanisms exhibited by insects, in general, serve to prevent invasion, growth, and processes of infectious pathogenic microorganism that could result in insect death.

### 1.3 Insect Hemolymph and Fungal Growth Medium Composition

During the entire infection process, all internal tissues of an insect, such as the gut, muscles, silk glands, and tracheae remain intact [7], suggesting that the cuticle and the nutrient-rich hemolymph are sufficient to complete the life cycle [1]. Fungal infection has been associated with the enzymes produced by the fungi [27]. Their growth [28] and production of hydrolytic enzymes depend on nutrients available. The nutritional requirement for the germination of fungal conidia on insect cuticle and the subsequent growth of hyphae has been established [29]. A minimum of at least three amino acids is necessary for a good germination and subsequent growth with the best combination being alanine, phenylalanine, and either leucine or valine. The absence of alanine always results in poor germination and hyphal outgrowth [29].

Trehalose is the principal sugar in the hemolymph of most insects [30, 31]. It is rapidly converted to glucose and used as substrate for glycolysis and the Krebs cycle [32]. Trehalose is present in 5–50 times higher concentrations than glucose [31] to make up for the inefficiency of the circulatory system in distributing nutrients by diffusion. There is, therefore, justification to use a high concentration of glucose in a submerged culture medium to grow entomopathogenic fungi. Mineral salt potassium has a stimulatory effect in the glycolytic cycle, while deficiency of phosphate may inhibit metabolism of sugar [33]. A deficit or excess of magnesium could influence the mycelia growth and metabolite production [33]. Total salts at 0.9–1.6% encompass the range of osmotic pressure in insect hemolymph [31, 32]. Insects in Dictyoptera largely use sodium and chloride, but with contributions from magnesium, potassium, and calcium [31]. Sodium is also an important cation in Diptera, Mecoptera, and Neuroptera but chloride is replaced by higher concentrations of amino acids [31]. In Hymenoptera and Lepidoptera, amino acids and other organic molecules play a major role along with potassium and lesser involvement of sodium [31]. These ionic differences were attributed to the insect diets. Plant-feeding insects contained higher levels of potassium and carnivorous insects had higher levels of sodium [31]. In the developing insect larva as well, most free amino acids are stored in the hemolymph which is in continuous contact with tissues and insect cells [34].

Analysis of free amino acids, derivatives, and peptides in 20 insect species belonging to seven orders—Lepidoptera, Diptera, Coleoptera, Orthoptera, Odonata, Hymenoptera, and Hemiptera—showed the presence of alanine, glutamate, glycine, leucine, proline, tyrosine, and valine in all the insects studied. During the development of the silkworm moth, the concentrations of methionine, glutamic acid, and aspartic acid are correlated with the activity of the silk glands [34]. Fungi require a nitrogen source to build their mass. A growth medium incorporated with bacteriological peptone, tryptone, and methionine can be used to provide an assorted range of amino acids that resembles the hemolymph of a broad range of insects and developing larvae. The ratio of the amino acids required, however, need not be fixed as the ratio in hemolymph fluctuates in insects depending on age, health, and food availability [34].

### 1.4 Fungal Enzymes

Fungi absorb simple nutrients like glucose directly but complex compounds need to be depolymerized prior to absorption. This is achieved by secreting extracellular enzymes. Entomopathogenic fungi are capable of producing a diversity of such lytic enzymes with multiple isoforms as observed in proteases (e.g., Pr1, Pr2, Pr3, Pr4) and chitinases [4, 16, 21, 35, 36], each playing their respective roles in the infection stages [9, 37, 38]. Degradation of the insect cuticle seems to be inefficient by either protease or chitinase activity alone but is enhanced when the different groups of enzymes act together or in sequence [9, 39]. The proteolytic enzymes secreted by the fungus expose the chitin structure concealed by the protein. Apart from hydrolyzing the cuticle and providing nutrient to the fungus, extracellular proteases may also hydrolyze antifungal proteins in the cuticle [21].

The subtilisin-like serine protease Pr1 produced by many entomopathogenic fungi plays a fundamental role in the infection process. It contributes to the degradation of cuticular protein at an early stage of fungal infection and during conidiation in the late stages of pathogenesis [40]. Protease Pr1 was found in high concentrations at the site of fungal penetration and its production was induced only by insect cuticle [21, 41, 42]. This enzyme seems to play a more important role in successful insect infection compared to others [21, 43]. When Pr1 was inhibited by a specific protease inhibitor from turkey egg white, melanization of the cuticle and invasion of the hemolymph was reduced, thus limiting fungal penetration and infection of the host [44]. Pr1 hydrolyses a broad range of substrates such as casein, elastin, bovine serum albumin, collagen, and insect cuticle proteins [35]. Pr1 cleaves at the C-terminal side of hydrophobic amino acid residues (e.g., Phe, Met, and Ala) [45] Any proteins with sequences containing $(Ala)_n$ repeats will be particularly susceptible as, e.g., the structural proteins from the cuticle of locust *Locusta migratoria* [46] and other insects which have similar sequences in at least some cuticle proteins.

Other proteinases involved in the infection process are Pr2, Pr3, and Pr4. Pr2 is a trypsin-type enzyme with a specificity for polypeptides containing the basic amino acids Lys and Arg. Pr2 degrades casein and albumin but not elastin [35]. Unlike Pr1, it is induced by a range of proteinaceous substrates. Pr4 is a cysteine protease that, like Pr2, attacks substrates on the C-terminal side of Arg and Lys [47]. Pr3 has an acidic pH optimum and is less studied [45]. Additional minor acidic proteases occur in some isolates of *M. anisopliae* [48]. Other deuteromycete entomopathogens reveal similar trends with each fungus producing multiple cuticle-degrading proteases in culture [48].

Fungi utilize chitinases to hydrolyze their own chitin and to break down the chitin of insects and other fungi. Digested polysaccharides consisting of β-(1,4) linked *N*-acetylglucosamine with a minimum chain length of $n=2$ [49] are absorbed as nutrients for growth, cell-wall remodeling. Fungal chitinases can be subdivided based on their amino acid composition with different substrate-binding site architectures [49]. Kim et al. [50] demonstrated the topical aphicidal activity of supernatant from a *B. bassiana* strain was strongly correlated to the fungal chitinase production [50] but not to its proteases Pr1 or Pr2 [51].

### 1.5 Measuring Enzyme Reactions

The speed and virulence success of an entomopathogenic fungus (infecting and eventually killing insects) will determine its practical use as a biological control agent in integrated pest management. Enzyme production and virulence can correlate and thus a fast, sensitive, and easy determination of chitinase and protease production in fungal cultures is useful to screen for virulent fungal candidates or simply to evaluate and understand the nutritional requirements and physiology of fungi grown under different growth parameters. Various qualitative and quantitative assessments [52] and reviews [53] on hydrolytic enzymes have been well developed, discussed, and often modified when deemed appropriate and necessary. Suitable substrates available for assays play an important role in measuring enzyme reaction in fungal cultures [52]. Some substrates, natural chitin, for instance, pose difficulties in enzyme reaction assessment. Synthetic substrates offer the advantages of increased sensitivity, true specificity, and ease of assay [52]. Both qualitative and quantitative assessments monitor hydrolytic enzymes breaking down substrates into simpler products with the ongoing enzyme reaction indicated via accumulation of the reaction products detectable via chromatography techniques, titration, or color change (chromogenic assay).

It is the latter technique using a synthetic substrate containing a fluorescent component or chromophore that makes visual assessment easy by eye (qualitative) or through use of an optical detector that determines the reaction intensity based on light emission or absorption (quantitative). A fluorimeter measures the fluorescence emitted from enzymatic reactions and a spectrophotometer measures the light absorbance by reaction product at a discrete wavelength in the visible and invisible spectrum [52]. Other enzymatic reactions have been measured by turbidimetry, viscometry, luminometry, and electrochemical methods [53]. All main factors (temperature, pH, ionic strength, and the proper concentrations of the essential components like substrates and enzymes) must be considered for assaying enzymes [52, 53].

With the advanced features of recent spectrophotometers, automated photometric assays allow (a) visualization of the enzymatic reaction over time at specific UV wavelength; (b) orbital shaking to mix enzyme reaction solutions; (c) maintenance of constant temperature, and the most convenient feature; and (d) the use of 96- or 384-well microtiter plates which can simultaneously accommodate large numbers of samples for enzyme reaction.

### 1.6 Exploiting the Potential of Chitinases and Proteases from Entomopathogenic Fungi

Microbial enzymes are preferred over enzymes from other biosources [54–56] for the development of industrial bioprocesses in pulp and paper, leather, detergents and textiles, pharmaceuticals, chemicals, food and beverages, biofuels, animal feed, cosmetics and general well-being, water treatment, agriculture, compost, fertilizer, among others. New and economically competitive enzymes with better or improved catalytic properties are continuously in demand driven by the need for sustainable applications. Entomopathogenic fungi could contribute to the production of different enzymes which fit different processes.

The following assay system describes a spectrophotometric approach utilizing a 96-well microtiter plate and specific substrates to measure the activities of chitinase and protease Prl in cell-free supernatants from a single isolate of the fungus *Beauveria bassiana* cultured for 7 days in three different media (glucose-, starch-, and insect material-based).

## 2 Materials

### 2.1 Chitinase Activity

1. 10 mM *p*-nitrophenyl-β-d-acetylglucosaminide (pNG powder) (Sigma-Aldrich).
2. 10 mM *p*-nitrophenol solution (Sigma-Aldrich).
3. 1 M potassium dihydrogen phosphate $KH_2PO_4$.
4. 1 M dipotassium hydrogen phosphate $K_2HPO_4$.
5. 0.1 M phosphate buffer (pH 6) can be prepared by mixing potassium dihydrogen phosphate ($KH_2PO_4$) and dipotassium monohydrogen phosphate ($K_2HPO_4$) (*see* **Note 1**) and diluting the combined 1 M stock solutions to 1 l with distilled water.
6. Chitinase from *Streptomyces griseus* (Sigma Aldrich).

### 2.2 Casein Gelatine Agar

1. 1% casein.
2. 1% gelatine.
3. 1.5% agar.
4. Cork borer.
5. Sterile petri dish.
6. Proteinase K (Sigma Aldrich).

### 2.3 Protease Pr1 Activity

1. 1 mM succinyl-alanine-alanine-2-proline-phenylalanine-*p*-nitroanilide (Synthetic peptide specific substrate) (Sigma Aldrich).
2. *p*-nitroaniline (Sigma Aldrich) prepared in ethanol (50 mg/ml) and serial diluted to prepare a standard curve.
3. 1 M Tris–HCl buffer: dissolve 121.1 g of Tris base in 800 ml distilled water and adjust pH to 8.5 with concentrated HCl. Make up to 1 l using distilled water.
4. Proteinase K (Sigma Aldrich).

## 3 Methods

The enzyme reaction in this approach was not terminated but monitored continuously. Concentrations of the reaction products that were obtained after substrate hydrolysis were estimated by using standard curves using the values from the optical density (OD) derived from different concentrations of commercial reaction product. Alternatively the enzyme activity of the cell-free supernatants could be compared directly using the OD differences (OD values from lapse time, $t_1$ – values from initial time, $t_0$) within the same interval.

### 3.1 Chitinase Activity

Chitinase activity is a measurement of the amount of *p*-nitrophenol released from *p*-nitrophenyl-β-d-acetylglucosaminide (pNG) substrate within a period of time [50]. The method was modified to allow for measurements using 96-well microtiter plates as the wells cannot hold over 300 μl solution.

1. A 50 μl cell-free supernatant sample was dispensed into a well containing 50 μl of 10 mM pNG substrate solution and 150 μl of 0.1 M phosphate buffer at pH 6 (*see* **Note 2**).
2. The samples were incubated at 37 °C and the kinetic assay was performed using a spectrophotometer (Thermo Scientific) at 405 nm.
3. At intervals of 30 min data collection was performed automatically to assess the chitinase activity in the supernatants.
4. The unit for chitinase activity was defined as the amount of enzyme required to release 1 mmol of *p*-nitrophenol per h per ml (*see* **Note 3**).
5. For comparison, the chitinase standard at 1 mg/ml in the buffer gave an average OD reading of 0.15 after 1 h incubation at 37 °C (Figs. 1 and 2).

### 3.2 Protease Activity

To determine the level of protease Pr1 present in the supernatants, the following method was used [57].

1. Each assay consisted of 50 μl of 1 mM succinyl-alanine-alanine-2-proline-phenylalanine-*p*-nitroanilide as a specific substrate, 0.85 ml of 15 mM Tris–HCl buffer (pH 8.5), and 0.1 ml cell-free fungal supernatant. The mixture was incubated for 1 h at 28 °C and upon hydrolysis the substrate released *p*-nitroaniline which was measured at 410 nm.
2. The data collection was performed at interval of 30 min. The amount of *p*-nitroaniline released was measured against its blank and standard curve at 410 nm (Fig. 3).

**Fig. 1** Chromogenic assay on 96-well microtiter plate reflecting enzyme reaction in supernatant samples

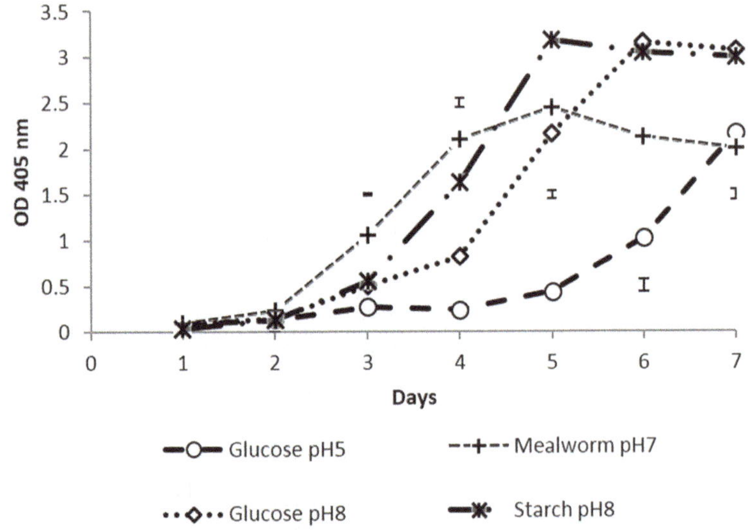


**Fig. 2** Chitinase activity after 1 h at 37 °C comparing the daily supernatant samples of the cultures grown in glucose-, starch-, and mealworm-based media. The error bars are the least significant difference (5 %)

3. Unit of enzyme activity (U) was defined as the amount of the enzyme able to release 1 nmol of *p*-nitroaniline (NA) per second per ml under the assay conditions (*see* **Note 4**).

### 3.3 Casein Gelatine Agar

1. Dissolved 1 % casein in 0.02 M NaOH and stirred until it formed a translucent solution.
2. Added 1 % gelatine and 1.5 % agar to the solution and adjusted pH to pH 7.0 with 1 M HCl. The media were sterilized and dispensed in 20 ml portions in sterile petri dishes. When the agar had solidified, wells were cut into the agar using a sterile cork borer (*see* **Note 5**).

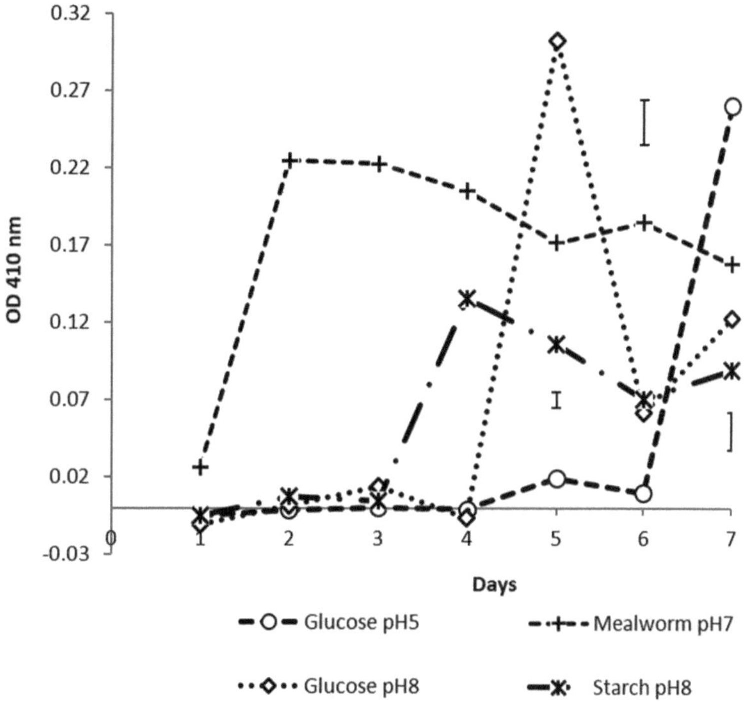


**Fig. 3** Pr1 activity comparing the daily supernatant samples of the cultures grown in glucose-, starch-, and mealworm-based media (activity after 1 h incubation at 37 °C). The error bars are the least significant difference (5 %)

3. 50 μl of each of the cell-free fungal supernatant samples was dispensed into a well in the agar. Proteinase K was included as positive control.
4. Plates were incubated at 37 °C for 24 h.
5. After 24 h of incubation, cloudy halos and distinct precipitation zones surrounding the wells could be detected in the casein-gelatin plates, suggesting proteolytic activity (*see* **Note 6**).

## 4 Notes

1. *See* Table 1. Modified from cshprotocols.org
2. Two replicates were carried out for all cell-free supernatant samples. Wells without fungal supernatant served as negative controls. Commercial chitinase from *Streptomyces griseous* (Sigma Aldrich) at 1 mg/ml was also included for assay as positive control.
3. Different concentrations of *p*-nitrophenol solution were dispensed in duplicate on the same plate. Absorbance data

**Table 1**
**Preparation of 0.1 M potassium phosphate buffer at room temperature 25 °C**

| pH | Volume of 1 M $K_2HPO_4$ (ml) | Volume of 1 M $KH_2PO_4$ (ml) |
|---|---|---|
| 5.8 | 8.5 | 91.5 |
| 6.0 | 13.2 | 86.8 |
| 6.2 | 19.2 | 8.08 |
| 6.4 | 27.8 | 72.2 |
| 6.6 | 38.1 | 61.9 |
| 6.8 | 49.7 | 50.3 |
| 7.0 | 61.5 | 38.5 |
| 7.2 | 71.7 | 28.3 |

obtained were used to plot a standard curve with absorbance at 410 nm against *p*-nitrophenol concentrations. The concentration of *p*-nitrophenol released from substrate using different cell-free supernatant samples with time was estimated from the standard curve. Alternatively the enzyme activity of the cell-free supernatants could be compared directly using the OD differences within the same interval.

4. On the same assay plate, different concentrations of *p*-nitroaniline (Sigma Aldrich) in DMSO or ethanol were prepared in duplicate. Absorbance data obtained at time zero were used to plot a standard curve with absorbance at 410 nm against *p*-nitroaniline (NA) concentrations. Proteinase K (Sigma Aldrich) was included as a positive control.
5. In addition, a dual substrate agar consisting of 1 % casein and 1 % gelatine [58] was also prepared to assess the production of the protease enzymes from the time course study.
6. Proteinase K, used as a standard, produced a distinct white opaque precipitation ring (Fig. 4). Daily supernatant samples (except 1-day-old sample) from a glucose-based *Beauveria* culture showed cloudy halos for all days (Fig. 4a). The 7-day-old sample showed distinct precipitation like proteinase K. The halos produced when starch-based *Beauveria* culture samples were applied were larger (Fig. 4b). All except 1-day-old daily samples from the mealworm-based *Beauveria* culture showed distinct precipitation (Fig. 4c).

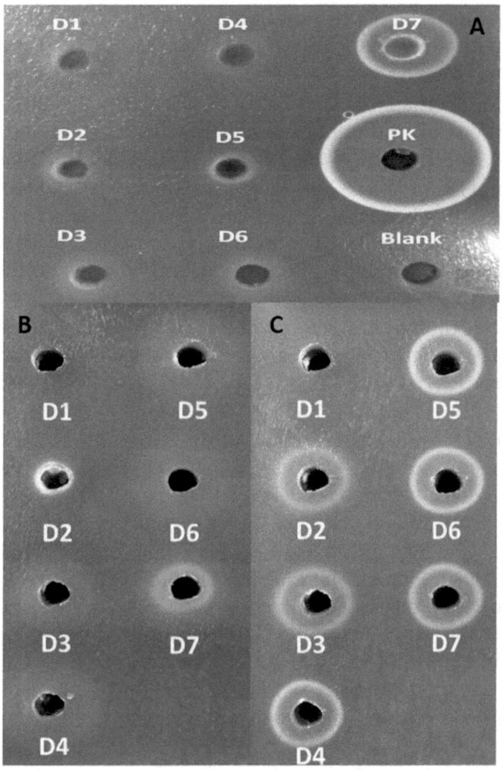


**Fig. 4** Casein-gelatin plates with daily supernatant samples added to wells. (**a**) Glucose-based supernatant, (**b**) starch-based supernatant, and (**c**) mealworm-based supernatant. *PK* Proteinase K

## References

1. Molnar I, Gibson DM, Krasnoff SB (2010) Secondary metabolites from entomopathogenic Hypocrealean fungi. Nat Prod Rep 27:1241–1275
2. Samson R, Evans HC, Latge JP (1988) Atlas of entomopathogenic fungi. Springer, Utrecht
3. Pekrul S, Grula EA (1979) Mode of infection of the corn earworm (Heliothis zea) by Beauveria bassiana as revealed by scanning electron microscopy. J Invertebr Pathol 34:238–247
4. Charnley AK (2003) Fungal pathogens of insects: cuticle degrading enzymes and toxins. In: Callow JA (ed) Advances in botanical research. Academic, San Diego, pp 241–321
5. Bidochka MJ, Khachatourians GG (1991) The implication of metabolic acids produced by Beauveria bassiana in pathogenesis of the migratory grasshopper, Melanoplus sanguinipes. J Invertebr Pathol 58:106–117
6. Butt T, Goettel MS (2000) Bioassays of entomogenous fungi. In: Navon A, Ascher KRS (eds) Bioassays of entomopathogenic microbes and nematodes. CABI Publishing, Wallingford, pp 141–196
7. Capinera JL (2008) Encyclopedia of entomology. Springer, New York
8. Goettel MS, Eilenberg J, Glare T (2010) Entomopathogenic fungi and their role in regulation of insect populations. In: Gilbert LI, Gill SS (eds) Insect control: biological and synthetic agents. Elsevier B.V., London
9. St. Leger RJ, Cooper RM, Charnley AK (1986) Cuticle-degrading enzymes of entomopathogenic fungi: cuticle degradation in vitro by enzymes from entomopathogens. J Invertebr Pathol 47:167–177
10. Inglis GD, Enkerli J, Goettel MS (2012) Chapter VII—Laboratory techniques used for entomopathogenic fungi: Hypocreales. In:

Lacey LA (ed) Manual of techniques in invertebrate pathology, 2nd edn. Academic, San Diego, pp 189–253

11. Juárez M, Fernández G (2007) Cuticular hydrocarbons of triatomines. Comp Biochem Physiol A Mol Integr Physiol 147:711–730
12. Hepburn HR (1985) Comprehensive insect physiology, biochemistry, and pharmacology. In: Kerkut GA, Gilbert LI (eds) Structure of the integument. Pergamon Press, New York
13. Bidochka M, Khachatourians G (1988) Regulation of extracellular protease in the entomopathogenic fungus Beauveria bassiana. Exp Mycol 12:161–168
14. Holder DJ, Kirkland BH, Lewis MW, Keyhani NO (2007) Surface characteristics of the entomopathogenic fungus Beauveria (Cordyceps) bassiana. Microbiology 153:3448–3457
15. Kirkland B, Westwood GS, Keyhani NO (2004) Pathogenicity of entomopathogenic fungi Beauveria bassiana and Metarhizium anisopliae to Ixodidae tick species Dermacentor variabilis, Rhipicephalus sanguineus and Ixodes scapularis. J Med Entomol 41:705–711
16. Ortiz-Urquiza A, Keyhani NO (2013) Action on the surface: entomopathogenic fungi versus the insect cuticle. Insects 4:357–374
17. Saito T, Aoki J (1983) Toxicity of free fatty acids on the larval surfaces of two lepidopterous insects towards Beauveria bassiana (Bals.) Vuill. and Paecilomyces fumoso-roseus (Wize) Brown et Smith (Deuteromycetes: Moniliales). Appl Entomol Zool 18:225–233
18. Smith RJ, Grula E (1982) Toxic components on the larval surface of the corn earworm (Heliothis zea) and their effects on germination and growth of Beauveria bassiana. J Invertebr Pathol 39:15–22
19. Sosa-Gomez D, Boucias D, Nation J (1997) Attachment of Metarhizium anisopliae to the southern green stink bug Nezara viridula cuticle and fungistatic effect of cuticular lipids and aldehydes. J Invertebr Pathol 69:31–39
20. Gołębiowski M, Cerkowniak M, Urbanek A, Dawgul M, Kamysz W, Boguś MI, Stepnowski P (2015) Identification and antifungal activity of novel organic compounds found in cuticular and internal lipids of medically important flies. Microbiol Res 170:213–222
21. Samuels RI, Paterson I (1995) Cuticle degrading proteases from insect moulting fluid and culture filtrates of entomopathogenic fungi. Comp Biochem Phys B 110:661–669
22. Samuels RI, Reynolds SE (2000) Proteinase inhibitors from the molting fluid of the pharate adult tobacco hornworm, Manduca sexta. Arch Insect Biochem Physiol 43:33–43
23. Bos N, Sundström L, Fuchs S, Freitak D (2015) Ants medicate to fight disease. Evolution 69:2979–2984
24. Rohlfs M, Churchill ACL (2011) Fungal secondary metabolites as modulators of interactions with insects and other arthropods. Fungal Genet Biol 48:23–34
25. Inglis GD, Johnson DL, Goettel MS (1996) Effects of temperature and thermoregulation on mycosis by Beauveria bassiana in grasshoppers. Biol Control 7:131–139
26. Chouvenc T, Su NY (2012) When subterranean termites challenge the rules of fungal epizootics. PLoS One 7(3):e34484
27. Roberts DW, Gupta S, St. Leger RJ (1992) Metabolite production of entomopathogenic fungi. Pesqui Agropecu Bras 27:325–347
28. Jaronski ST, Jackson MA (2012) Chapter VIII—Mass production of entomopathogenic Hypocreales. In: Lacey LA (ed) Manual of techniques in invertebrate pathology, 2nd edn. Academic, San Diego, pp 255–284
29. Smith RJ, Grula EA (1981) Nutritional requirements for conidial germination and hyphal growth of Beauveria bassiana. J Invertebr Pathol 37:222–230
30. Harrison JF, Woods HA, Roberts SP (2012) Ecological and environmental physiology of insects. Oxford University Press, Oxford
31. Klowden MJ (2013) Physiological systems in insects. Elsevier Science, Amsterdam
32. Nation JL (2008) Insect physiology and biochemistry, 2nd edn. CRC Press, Boca Raton
33. Wang P, Jiang X, Jiang Y, Hu X, Hwang H (2008) Optimization of fermentation medium and conditions for mycelial growth and water-soluble exo-polysaccharides production by Isaria farinosa B05. Prep Biochem Biotechnol 38:294–307
34. Chen P (2013) Amino acid and protein metabolism. In: Kerkut GA (ed) Biochemistry. Elsevier Science, Amsterdam, pp 117–218
35. Leger RJ, Cooper RM, Charnley AK (1987) Production of cuticle-degrading enzymes by the entomopathogen Metarhizium anisopliae during infection of cuticles from Calliphora vomitoria and Manduca sexta. J Gen Microbiol 133:1371–1382
36. Kučera M (1971) Toxins of the entomophagous fungus Beauveria bassiana: II. Effect of nitrogen sources on formation of the toxic protease in submerged culture. J Invertebr Pathol 17:211–215
37. St. Leger RJ, Frank DC, Roberts DW, Staples RC (1992) Molecular cloning and regulatory analysis of the cuticle-degrading-protease structural gene from the entomopathogenic

fungus Metarhizium anisopliae. Eur J Biochem 204:991–1001
38. St. Leger RJ, Joshi L, Roberts D (1998) Ambient pH is a major determinant in the expression of cuticle-degrading enzymes and hydrophobin by Metarhizium anisopliae. Appl Environ Microbiol 64:709–713
39. Fang W, Feng J, Fan Y, Zhang Y, Bidochka MJ, Leger RJS, Pei Y (2009) Expressing a fusion protein with protease and chitinase activities increases the virulence of the insect pathogen Beauveria bassiana. J Invertebr Pathol 102:155–159
40. Small CLN, Bidochka MJ (2005) Up-regulation of Pr1, a subtilisin-like protease, during conidiation in the insect pathogen Metarhizium anisopliae. Mycol Res 109:307–313
41. St. Leger RJ, Butt TM, Staples RC, Roberts DW (1989) Synthesis of proteins including a cuticle-degrading protease during differentiation of the entomopathogenic fungus Metarhizium anisopliae. Exp Mycol 13:253–262
42. Goettel MS, Leger RJS, Rizzo NW, Staples RC, Roberts DW (1989) Ultrastructural localization of a cuticle-degrading protease produced by the entomopathogenic fungus Metarhizium anisopliae during penetration of host (Manduca sexto) cuticle. Microbiology 135:2233–2239
43. Reynolds SE, Samuels RI (1996) Physiology and biochemistry of insect moulting fluid. Adv Insect Physiol 26:157–232
44. St. Leger RJ, Durrands PK, Charnley AK, Cooper RM (1988) Role of extracellular chymoelastase in the virulence of Metarhizium anisopliae for Manduca sexta. J Invertebr Pathol 52:285–293
45. St. Leger RJ, Charnley AK, Cooper RM (1987) Characterization of cuticle-degrading proteases produced by the entomopathogen Metarhizium anisopliae. Arch Biochem Biophys 253:221–232
46. Andersen SO, Hojrup P, Roepstorff P (1995) Insect cuticular proteins. Insect Biochem Mol Biol 25:153–176
47. Cole SCJ, Charnley AK, Cooper RM (1993) Purification and partial characterisation of a novel trypsin-like cysteine protease from Metarhizium anisopliae. FEMS Microbiol Lett 113:189–195
48. St. Leger RJ, Cooper RM, Charnley A (1987) Distribution of chymoelastases and trypsin-like enzymes in five species of entomopathogenic deuteromycetes. Arch Biochem Biophys 258:123–131
49. Hartl L, Zach S, Seidl-Seiboth V (2012) Fungal chitinases: diversity, mechanistic properties and biotechnological potential. Appl Microbiol Biotechnol 93:533–543
50. Kim JS, Roh JY, Choi JY, Wang Y, Shim HJ, Je YH (2010) Correlation of the aphicidal activity of Beauveria bassiana SFB-205 supernatant with enzymes. Fungal Biol 114:120–128
51. Kim JS, Je YH, Woo EO (2010) Roles of adjuvants in aphicidal activity of enzymes from Beauveria bassiana (Ascomycota: Hypocreales) SFB-205 supernatant. J Asia Pacific Entomol 13:345–350
52. Kramer KJ, Dziadik-Turner C, Koga D (2013) Chitin metabolism in insects. In: Kerkut GA (ed) Integument, respiration and circulation. Elsevier Science, Amsterdam
53. Bisswanger H (2014) Enzyme assays. Perspect Sci 1:41–55
54. Barredo JL (2005) Microbial enzymes and biotransformations. Humana Press, New York
55. Anbu P, Gopinath SCB, Chaulagain BP, Tang TH, Citartan M (2015) Microbial enzymes and their applications in industries and medicine 2014. BioMed Research International Article ID 816419
56. Gurung N, Ray S, Bose S, Rai V (2013) A broader view: microbial enzymes and their relevance in industries, medicine, and beyond. BioMed Res Int 2013:329121
57. Donatti AC, Furlaneto-Maia L, Fungaro MHP, Furlaneto MC (2008) Production and regulation of cuticle-degrading proteases from Beauveria bassiana in the presence of Rhammatocerus schistocercoides cuticle. Curr Microbiol 56:256–260
58. Montville T (1983) Dual-substrate plate diffusion assay for proteases. Appl Environ Microbiol 45:200–204

# Chapter 15

# Analytical Methods for Secondary Metabolite Detection

**Judith Taibon and Hermann Strasser**

## Abstract

The entomopathogenic fungi *Metarhizium brunneum*, *Beauveria bassiana*, and *B. brongniartii* are widely applied as biological pest control agent in OECD countries. Consequently, their use has to be flanked by a risk management approach, which includes the need to monitor the fate of their relevant toxic metabolites. There are still data gaps claimed by regulatory authorities pending on their identification and quantification of relevant toxins or secondary metabolites. In this chapter, analytical methods are presented allowing the qualitative and quantitative analysis of the relevant toxic *B. brongniartii* metabolite oosporein and the three *M. brunneum* relevant destruxin (dtx) derivatives dtx A, dtx B, and dtx E.

**Key words** Secondary metabolite, Toxin, Oosporein, Destruxin, Persistence, Risk assessment, Analyte monitoring, Analytical tool, HPLC–DAD, HPLC–DAD–QTOF–MS/MS

## 1 Introduction

The regulatory authorities worldwide want to have a stringent procedure for the assessment of potentially toxic metabolic by-products by candidate micro-organisms which are used as pest control products. The majority of fungal biocontrol agents (BCAs) are based on the mitosporic ascomycetes *Beauveria bassiana*, *B. brongniartii* or one of the type species of *Metarhizium* spp. These fungi have a number of advantages as biocontrol agents. They are easily mass produced, strains can be selected with appropriate levels of virulence and specificity, and formulations with increased shelf life and field efficacy are possible. Nevertheless, risks and hazard linked to the production of metabolites and toxins by these anamorphic entomopathogenic fungi are still under discussion. Data gaps have been reported by European Food Safety Authority (EFSA) in their conclusion on the peer review of the pesticide risk assessment of the two active substances, i.e., *M. brunneum* BIPESCO 5/F52 (formerly *M. anisopliae* var. *anisopliae*) and *B. bassiana* GHA (EFSA J. 2012 and EFSA J. 2013). EFSA claimed that no "information has been provided on the production and

Travis R. Glare and Maria E. Moran-Diez (eds.), *Microbial-Based Biopesticides: Methods and Protocols*, Methods in Molecular Biology, vol. 1477, DOI 10.1007/978-1-4939-6367-6_15, © Springer Science+Business Media New York 2016

persistence of metabolites" produced by these fungi and that "the consumer risk assessment could not be finalized until the issue of toxins/secondary metabolites" has been addressed. Further information is requested to assess the production and fate behavior of relevant toxic metabolites such as oosporein and destruxins.

According to this important request expert groups intensified their analytical work assessing these toxic relevant metabolites in selected matrices. Seger and coauthors published a validated high-performance liquid chromatography–diode array detection assay (HPLC–DAD) for the detection and quantification of the *Beauveria* metabolite oosporein from fungal culture broth and two biocontrol formulations (Melocont™-Pilzgerste and Melocont™-WP) as well as from potato tubers [1, 2]. For destruxin isolation [3] and analysis different chromatographic assays such as HPLC methods [4–13], one assay performing capillary electrophoresis (CE) [14] and one previously published supercritical fluid chromatography-based assay (SFC) [15] are reported in literature.

In this chapter, analytical methods are presented allowing the qualitative and quantitative analysis of the relevant toxic *B. brongniartii* metabolite oosporein (*see* Fig. 1) and the three *M. brunneum* relevant destruxin (dtx) derivatives dtx A, dtx B, and dtx E (*see* Fig. 2).

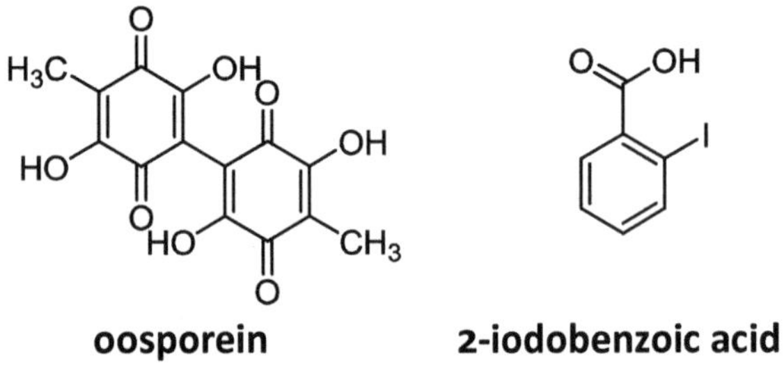


**Fig. 1** Structure of oosporein and 2-iodobenzoic acid

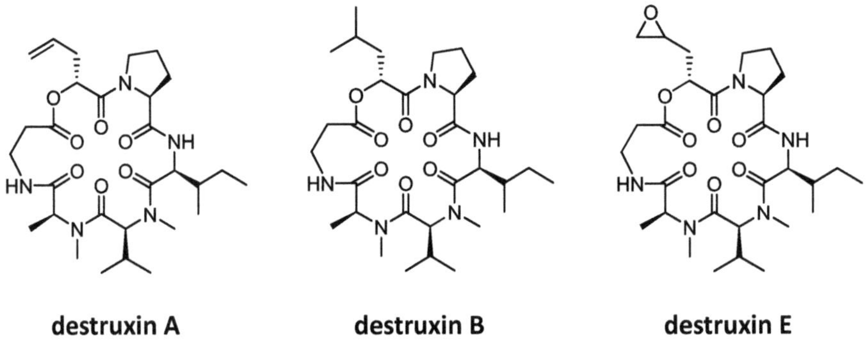


**Fig. 2** Structural formulae of the main destruxins A, B, and E

## 2 Materials

### 2.1 *Beauveria brongniartii*

All solutions were prepared at room temperature using HPLC water produced by reversed osmosis followed by distillation and analytical grade reagents.

#### 2.1.1 Materials for Submerged Culture

1. Solution for inoculum: Sterile 0.05 % (w/v) aqueous Tween 80 solution.
2. Liquid medium (S2G) I: Dissolve 30 g of SABOURAUD-2 %-Glucose Bouillon with deionized water in a 1 L Erlenmeyer flask.
3. Liquid medium (S2G) II: S2G medium (I) containing 0.05 % (v/v) Silicon 289 antifoam (Sigma).
4. 3 L Erlenmeyer flasks.
5. Gyratory shaker (Certomat® RO, Braun, shaking amplitude 50 mm).
6. Stirred tank reactor (NLF22, Bio Engineering).
7. Cotton cloth.
8. Filter paper (0.7 μm, Machery Nagel 615).

#### 2.1.2 Fungal Growth on Barley Kernels

1. Barley kernels: Crush the sample material retaining the husk.
2. Polypropylene bags with paper clips.

#### 2.1.3 Buffer Preparation

1. Buffer stock solutions: Prepare stock solutions of 100 g/L for acetic acid, phosphoric acid, and sodium hydroxide and 10 g/L for boric acid.
2. Britton–Robinson buffer (pH = 5.5): Mix 5.13 mL of stock solution acetic acid, 8.40 mL of stock solution phosphoric acid, 53.1 mL of stock solution boric acid, and 6.84 mL of stock solution sodium hydroxide and add distilled water to a final volume of 300 mL (*see* **Note 1**).
3. BR5.5-MeOH sample buffer: Prepare the sample buffer by diluting the Britton–Robinson buffer with methanol (3:7 v/v) (*see* **Note 2**).
4. Methanol.
5. pH meter.

#### 2.1.4 Extraction of Melocont™-WP, Melocont™-Pilzgerste, and Biological Samples

1. Melocont™-WP from AgResearch, Lincoln, New Zealand.
2. Melocont™-Pilzgerste from F. Joh. Kwizda GmbH, Austria.
3. Potato tubers (*Solanum tuberosum*) obtained from field trials where *B. brongniartii* was used as BCA.
4. Kitchen blender.
5. Centrifuge (Heraeus Labofuge 400, Swinging bucket rotor, radius 11.3 cm).

6. 5 mL volumetric flask.
7. Ultrasonic bath.
8. Vivaspin2 (Sartorius) (*see* **Note 3**).
9. Polypropylene tubes.
10. HPLC vials.

*2.1.5 Preparation of Stock Solutions and Solvents*

1. Oosporein stock solution: prepare an oosporein stock solution in BR5.5-MeOH buffer.
2. 2-iodobenzoic acid stock solution (*see* Fig. 1): Prepare an internal standard stock solution (IS) by exact weighing of 2-iodobenzoic acid (analytical grade, Sigma-Aldrich) in methanol (*see* **Note 4**).
3. Extraction solvent: add an appropriate amount of the IS stock solution to the BR5.5-MeOH sample buffer.
4. Mobile phase A: water containing 0.1 % (v/v) acetic acid and 0.9 % (v/v) formic acid.
5. Mobile phase B: acetonitrile containing 0.1 % (v/v) acetic acid and 0.9 % (v/v) formic acid.

*2.1.6 HPLC–DAD Conditions*

1. HP 1100 liquid chromatograph (Agilent) equipped with a diode array detector (DAD), an automatic injector, an autosampler, and a column oven.
2. Phenomenex Synergi Hydro-RP 80A column (150 × 2 mm), particle size 4 μm (Phenomenex).
3. Security Guard system equipped with a C18 cartridge (4 × 2 mm) (Phenomenex).

*2.1.7 HPLC-DAD Assay Validation: Submerged Culture Broth and Biocontrol Formulations*

1. Use an oosporein stock solution to prepare a dilution series in BR5.5-MeOH sample buffer.
2. Prepare for intraday and interday repeatability three different fortification levels of oosporein in BR5.5-MeOH sample buffer.
3. Recovery experiments: Prepare a BR5.5-MeOH sample buffer solution fortified with different concentrations of oosporein. The concentration should be adjusted to the oosporein concentration found in the respective matrix.

*2.1.8 HPLC–DAD Assay Validation: Potato Tubers*

1. Use an oosporein stock solution to prepare a dilution series in BR5.5-MeOH sample buffer.
2. Use an IS stock solution to prepare calibration levels in BR5.5-MeOH sample buffer.
3. Prepare solutions containing oosporein and IS by adding an aliquot of the IS stock solution to the oosporein dilution series.

4. Use potato tuber aliquots (e.g. 40 g) from untreated field plots to prepare calibration levels for the quantitative assay. Prepare for sample preparation the extraction solvent (BR5.5-MeOH sample buffer containing the IS) fortified with different concentrations of oosporein.
5. Use for recovery experiments potato tuber aliquots and the extraction solvent (BR5.5-MeOH sample buffer containing the IS) fortified with three different concentrations of oosporein.

## 2.2 *Metarhizium brunneum*

All solutions were prepared at room temperature using HPLC water produced by reversed osmosis followed by distillation and analytical grade reagents.

### 2.2.1 Cultivation of *Metarhizium brunneum*

1. SABOURAUD-4 %-Glucose agar plates: dissolve 65 g of SABOURAUD-4 %-Glucose agar with deionized water in a 1 L Erlenmeyer flask.
2. Use single spore isolates of *M. brunneum* to inoculate SABOURAUD-4 %-Glucose agar plates.
3. Preculture: Prepare 200 mL of a spore suspension in a sterile 0.1 % (w/v) aqueous Tween 80 solution.
4. Liquidmedium(S4G)I:Dissolve50gofSABOURAUD-4%-Glucose Bouillon with deionized water in a 1 L Erlenmeyer flask.
5. Liquid medium (S4G) II: Prepare 7.8 L of the S4G liquid medium I and add 4 mL of antifoam agent (Clerol FBA 5075).
6. 500 mL Erlenmeyer flasks.
7. Gyratory shaker (Certomat® RO, Braun, shaking amplitude 50 mm).
8. Stirred tank reactor (NLF Bioengineering D304).
9. Cotton cloth.
10. Filter paper (0.7 μm, Machery Nagel 615).

### 2.2.2 Extract Preparation

1. Extraction solvent: dichloromethane.
2. Reagents: water and sodium sulfate.
3. Separation funnel.
4. Filter paper (0.7 μm, Machery Nagel 615).
5. Rotavapor.

### 2.2.3 Thin Layer Chromatography for Fraction Analysis

1. TLC aluminum plates (Silica gel 60 $F_{254}$, Merck).
2. Mobile phase: Mix ethyl acetate/acetone/water/formic acid at a ratio of 60:40:2:1 (*see* **Note 5**).
3. Twin trough chamber (CAMAG).
4. Derivatization reagent: Prepare a 1 % ethanolic iodine solution (w/v).
5. Hairdryer.

#### 2.2.4 HPLC–DAD and HPLC–DAD–MS/MS Conditions for Fraction Analysis

1. HP 1100 liquid chromatograph (Agilent) equipped with a diode array detector (DAD), an automatic injector, an autosampler, and a column oven.
2. Esquire 3000$^{plus}$ ion-trap mass spectrometer (Bruker Daltonics).
3. Zorbax SB-C18 column (150×2 mm), particle size 3.5 μm (Agilent Technologies).
4. Guard column: A LiChroCART 4-4 (Merck) column (4 mm×4 mm) filled with LiChrospher 100 RP-18 material (particle size 5 μm).
5. Mobile phase: water (A) and acetonitrile (B).

#### 2.2.5 Chromatography on Sephadex LH-20 Material

1. Crude extract of *Metarhizium brunneum*.
2. Organic solvents for chromatography on Sephadex LH-20: dichloromethane, acetone, and mixtures at different ratios (dichloromethane:acetone 85:15 (v/v), dichloromethane:acetone 1:1 (v/v)).
3. Stationary phase: 150 g Sephadex LH-20 (Sigma Aldrich).
4. Column: 100 cm×4 cm (Kronlab).
5. Test tubes with a volume of >6 mL.
6. Fraction collector (SuperFrac, Pharmacia Biotech).
7. Preweight glasses.
8. SpeedVac Plus SC 210A (Thermo Savant).

#### 2.2.6 High-Speed Counter-Current Chromatography (HSCCC)

1. Preparative HSCCC Model CCC 1000 multilayer coil counter-current chromatograph equipped with a 325 mL coil column and an electronic controller (Pharma-Tech-Research).
2. HPLC pump (LC-10AD-VP, Shimadzu).
3. Manual sample injection valve with a 10 mL sample loop.
4. Components of the solvent system: Mix light petroleum:ethyl acetate:methanol:water at a ratio of 2:5:2:5 and equilibrate the mixture by repeatedly shaking in a separation funnel at room temperature. Degas separated phases for 10 min in an ultrasonic bath before use. Use the lower phase as mobile phase and the upper phase as stationary phase (*see* **Note 6**).
5. Test tubes with a volume of >10 mL.
6. Fraction collector (SuperFrac, Pharmacia Biotech).
7. Preweight glasses.
8. SpeedVac Plus SC 210A (Thermo Savant).

#### 2.2.7 Sample Preparation from Culture Filtrate

1. Prepare samples by solid phase extraction (SPE) using Strata C18-E cartridges (100 mg/1 mL, Phenomenex).
2. Solvents: methanol, 85% methanol (v/v), 40% methanol (v/v), and water.

3. HPLC Vials.
4. Centrifuge (Heraeus Labofuge 400, Swinging bucket rotor, radius 11.3 cm).

*2.2.8 UHPLC-DAD-QTOF-MS/MS Conditions*

1. 1200 UHPLC liquid chromatograph (Agilent) equipped with a diode array detector (DAD), an automatic injector, an autosampler, and a column oven.
2. Zorbax Eclipse XDB-C18 rapid resolution column (50×2.1 mm, 1.8 μm particle size), guarded with an inline filter (0.2 μm pore size frit, 2.1 mm diameter).
3. Mobile phase: water (A) and acetonitrile (B), each containing 0.02 % acetic acid (v/v).
4. Bruker micrOTOF-QII mass spectrometer (Bruker Daltonics).

*2.2.9 Assay Validation*

1. Prepare a stock solution containing dtx A, B, and E by exact weighing the analytes and dissolving them in methanol.
2. Prepare outgoing of these stock solutions spike solutions containing dtx A, B, and E.
3. Prepare final calibration levels by a 1:10 dilution of each spike solution into either methanol or culture medium (*see* **Note** 7).
4. Prepare a second methanolic stock solution for recovery experiments. Generate four spike solutions (covering the range of the calibration curve) and obtain samples by a 1:10 dilution of each spike solution into culture medium. Prepare all levels in three replicates.

## 3 Methods

### 3.1 *Beauveria brongniartii*

*3.1.1 Submerged Culture of B. brongniartii*

1. Suspend conidia from 14- to 20-day-old cultures in the sterile 0.05 % (w/v) aqueous Tween 80 solution and inoculate a 3 L Erlenmeyer flask containing 1.5 L S2G liquid medium I to a final concentration of $1 \times 10^6$ conidia/mL (*see* **Note 8**).
2. Incubate cultures for 6 days on a gyratory shaker at 25 °C and 150 rpm.
3. Use this culture to inoculate a 14 L stirred tank reactor (Bio Engineering NLF22) containing S2G liquid medium II (*see* **Note 9**). Final concentration should be 10 % (v/v).
4. Incubate the culture for 4 days at 23 °C and 350–400 rpm to ensure a dry biomass >7 g/L.
5. Culture filtrate: Separate the mycelium by filtration through cotton cloth. The final purification of the culture broth is done by filtering through filter paper (*see* **Note 10**).
6. Store the culture filtrate at −20 °C until needed.

*3.1.2 Fungal Growth on Barley Kernels*

1. Place 2.0 kg of the prepared barely kernels into polypropylene bags, add water (60% w/v) and autoclave it twice for 1.5 h at 121 °C.
2. Inoculate each autoclaved bag with 100 mL culture broth and mix everything before sealing with paper clips (*see* Subheading 3.1.1, **steps 1–4**).
3. Mix the contents again after 5 days incubation at 25 °C (*see* **Note 11**) and incubate them for another 9 days. Store the fully colonized kernels (after 14 days) at 4 °C until required.

*3.1.3 Extraction from Culture Filtrate*

1. Dilute 50 μL of a culture filtrate sample (*see* Subheading 3.1.1, **steps 5** and **6**) with 950 μL of sample buffer (BR5.5-MeOH) and transfer it into a HPLC vial. The sample can be measured without further purification.

*3.1.4 Extraction from Barley Kernels or Melocont™-Pilzgerste*

1. Submerse 10 g of frozen (−20 °C) colonized barley kernels or Melocont™-Pilzgerste in 200 mL of sample buffer (BR5.5-MeOH) and mill all in a kitchen blender (*see* **Note 12**).
2. Centrifuge the sample for 30 min at 1547 × *g* (room temperature) to pellet matrix components.
3. Transfer an aliquot of 1 mL of the clear supernatant to a HPLC vial for measurement.

*3.1.5 Extraction from Melocont™-WP*

1. Suspend 0.1 g of Melocont™-WP in a 5 mL volumetric flask using sample buffer (BR5.5-MeOH).
2. Place the volumetric flask in an ultrasonic bath and sonicate three times for 5 min (*see* **Note 13**).
3. Purify samples by centrifugation (20 min at 789 × *g*, room temperature) of 2 mL over an acetylated cellulose (CTA) membrane (Vivaspin2).
4. Transfer 1 mL of the ultrafiltrate to HPLC vials for measurement.

*3.1.6 Extraction of Biological Samples: Potato Tubers*

1. Use for sample preparation the prepared extraction solvent (*see* **Note 14**).
2. Chop frozen potato tubers of 200–400 g weight into pieces of 5–20 g. Take aliquots of 35–45 g, transfer them into a kitchen blender and mix all with 200 mL extraction solvent (BR5.5-MeOH sample buffer containing the IS). Blend the sample material for 5 min (*see* **Note 15**).
3. Transfer 10 mL of the blended sample to polypropylene tubes and centrifuge them for 30 min at 3500 rpm (room temperature).
4. Transfer 1 mL of the clear supernatant to HPLC vials for measurement.

*3.1.7 HPLC–DAD Assay for the Analysis and Quantification of Oosporein*

1. Perform the separation of analytes on a Phenomenex Synergi Hydro-RP column equipped with a C18 cartridge used as guard column.
2. Run the analysis using a gradient method with the solvents water (A) and acetonitrile (B) both containing 0.1 % (v/v) acetic acid and 0.9 % (v/v) formic acid (*see* **Note 16**). The course of the gradient is 5–60 % B in 6 min, followed by 60–98 % B in 2 min, keeping it constant for further 5 min. The flow rate is set at 0.3 mL/min and the oven temperature at 23 °C.
3. Reequilibrate the column between single runs for 7 min using starting conditions.
4. Set the injection volume at 2 μL and record the chromatograms at 287 nm.
5. Run all samples with these HPLC conditions (*see* Figs. 3 and 4) (*see* **Note 17**).

*3.1.8 HPLC–DAD Assay Validation: Submerged Culture Broth and Biocontrol Formulations*

1. Use a dilution series of oosporein in BR5.5-MeOH sample buffer to obtain calibration curves. Measure all calibration levels three times.
2. Use for intraday and interday repeatability three fortification levels of oosporein in BR5.5-MeOH sample buffer. Measure all samples three times on 3 different days.
3. Perform recovery experiments from different matrices using BR5.5-MeOH sample buffer fortified with oosporein (*see* **Note 18**). Prepare samples in triplicates and perform sample preparations as described for the respective matrix.
4. Assess from this experimental setup the validation parameters LOD, LOQ, precision, accuracy, intra- and interday repeatability of the method.

*3.1.9 HPLC–DAD Assay Validation: Potato Tubers*

1. Prepare calibration curves to assess the linearity of IS in BR5.5-MeOH sample buffer and of oosporein and IS in the presence of each other. Measure all levels of the IS and oosporein dilution series three times.
2. Prepare calibration levels from potato tubers as described and measure all samples three times.
3. Evaluate analyte and IS loss during sample preparation performing recovery experiments. Prepare samples as described and measure all three times.
4. Assess from this experimental setup the validation parameters LOD, LOQ, precision, accuracy, intra- and interday repeatability of the method.

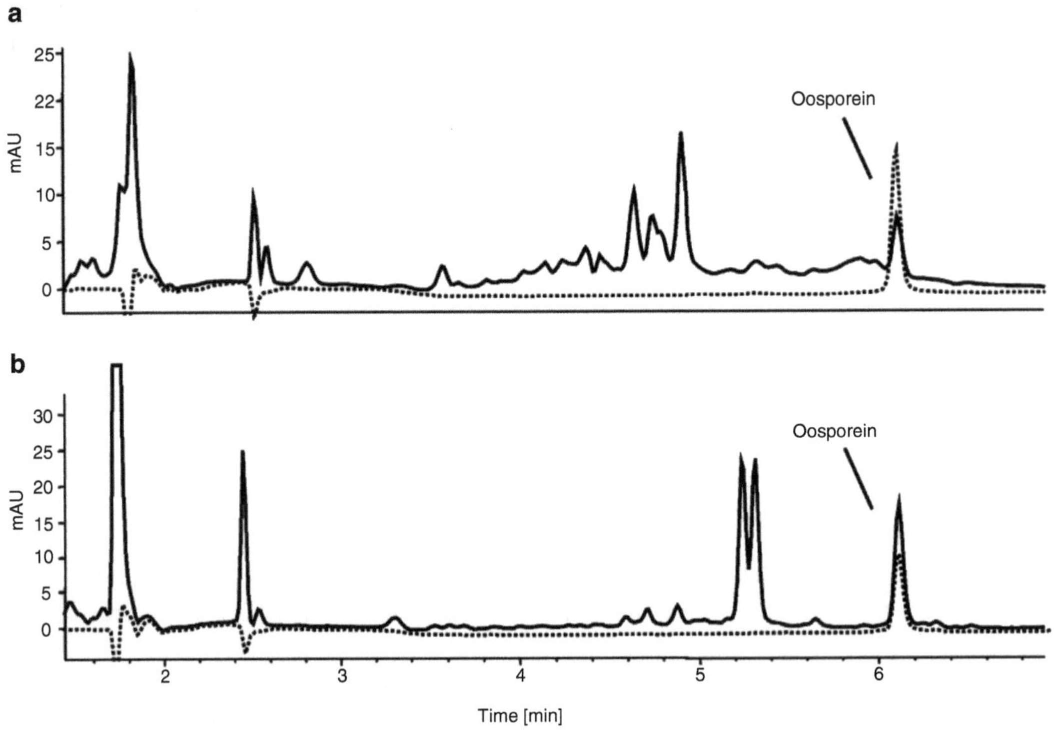


**Fig. 3** Representative HPLC–DAD chromatograms: (**a**) Oosporein content in Melocont™-Pilzgerste. *Full line*: chromatogram of a Melocont™-Pilzgerste sample; *Dashed line*: oosporein solution in BR5.5-MeOH sample buffer used as extraction solvent in the spiking experiment. (**b**) Oosporein in Melocont™-WP: *Full line*: chromatogram of a Melocont™-WP sample; *Dashed line*: oosporein solution in BR5.5-MeOH sample buffer used as extraction solvent in the spiking experiment. Concentrations of oosporein in the extraction solvent used are equivalent to concentrations found in the respective matrix. Reproduced from Seger et al. [1] with permission from American Chemical Society

### 3.2 *Metarhizium brunneum*

#### 3.2.1 *Cultivation of Metarhizium brunneum*

1. Inoculate SABOURAUD-4 %-Glucose agar plates with single spore isolates and incubate plates for 2 weeks at 25 °C and a relative humidity of 65 %.
2. Prepare 200 mL of a preculture for inoculation of the stirred tank reactor. Use the prepared spore suspension to inoculate 500 mL Erlenmeyer flasks containing 50 mL of S4G liquid medium I. The concentration of conidia should be $2.5 \times 10^7$ conidia per flask. Incubate the flasks on a gyratory shaker for 5 days at 200 rpm, 25 °C, and a relative humidity of 65 %. Replace the liquid lost due to evaporation with deionized water.
3. Use this preculture to inoculate a 10 L stirred tank reactor containing 7.8 L of S4G liquid medium II (*see* **Note 19**).
4. Incubate the culture for 21 h at 300 rpm, increase to 350 rpm for 2 h, and adjust to 400 rpm until the end of incubation at 70 h is reached. Operate the reactor during incubation at a

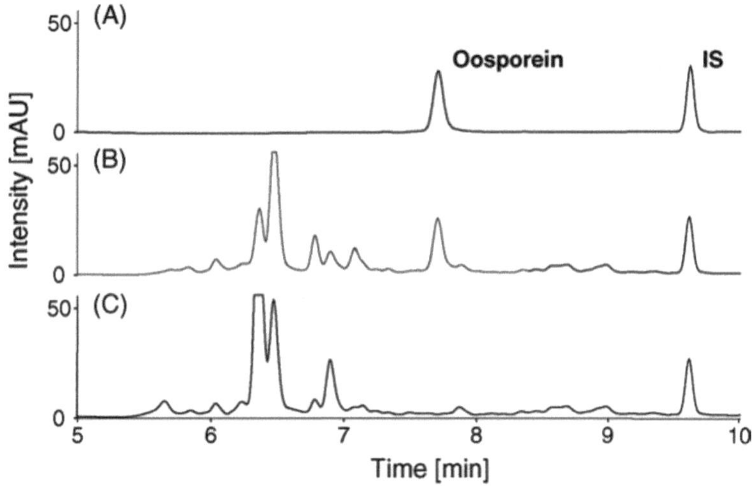


**Fig. 4** Representative HPLC-DAD chromatograms recorded at 287 nm: (**a**) Sample buffer (BR5.5-MeOH) spiked with oosporein and 2-iodobenzoic acid; (**b**) Potato tuber sample from an untreated field plot extracted with 200 mL extraction solvent (BR5.5-MeOH sample buffer containing IS) fortified with oosporein; (**c**) Potato tuber sample from a field plot treated with Melocont™-Pilzgerste extracted with 200 mL extraction solvent. Reproduced from Seger et al. [2] with permission from Elsevier

temperature of 25 °C and ventilate with 1 volume air/volume liquid/minute (vvm).

5. Replace the liquid lost with deionized water before harvesting the culture broth.
6. Separate the mycelium by filtration through cotton cloth. The final purification of the culture broth is done by filtering through filter paper (*see* **Note 10**).
7. Store the culture filtrate at –20 °C until needed.

*3.2.2 Preparation of Crude Extract*

1. Extract aliquots of the culture filtrate five times with dichloromethane at a ratio of 3 volumes/volume culture filtrate (*see* **Note 20**). Perform all preparative operations with organic solvents under a well-vented hood to minimize the impact on the laboratory staff.
2. Wash combined organic layers twice with 50 mL water (*see* **Note 21**) and dry them with sodium sulfate (*see* **Note 22**).
3. Remove the sodium sulfate by filtration and evaporate the organic layer to dryness using a rotavapor (*see* **Note 23**).
4. Dried crude extracts were stored at –20 °C until needed.

*3.2.3 Chromatography on Sephadex LH-20 Material*

1. Prepare the stationary phase bed by equilibrating 150 g of Sephadex LH-20 material overnight in dichloromethane:acetone 85:15 (v/v) (*see* **Note 24**).

2. Transfer the slurry to the column (100 cm × 4 cm) and wash the bed after it was settled down with 1 L of dichloromethane:acetone 1:1 (v/v).
3. Equilibrate the column with 1 L of dichloromethane prior to separation.
4. Apply 3.5 g of the crude extract (*see* Subheading 3.2.2) dissolved in 15 mL of dichloromethane to the column (*see* **Note 25**). Try to apply the sample as concentrated as possible to achieve a tight start zone.
5. Elution is carried out with 1000 mL of dichloromethane followed by 500 mL of dichloromethane: acetone 85:15 (v/v), 500 mL of dichloromethane:acetone 1:1 (v/v), and 500 mL of acetone at a flow rate of 1.2 mL/min.
6. Collect the eluate into test tubes at a fraction size of 6 mL using a fraction collector.
7. Analyze fractions either by thin layer chromatography (*see* Subheading 3.2.5) or by HPLC–DAD–MS/MS described in Subheading 3.2.6 (*see* **Note 26**). Combine fractions with same content, transfer them to preweight glasses, and evaporate them to dryness using the SpeedVac Plus.

#### *3.2.4 High-Speed Counter-Current Chromatography (HSCCC)*

1. Use this technique for final purification of analyte enriched fractions obtained by column chromatography over Sephadex material (*see* Subheading 3.2.3).
2. Prepare sample solution by dissolving the sample in ethyl acetate (800 mg/9 mL).
3. Choose the right operation mode of the system (*see* **Note 27**).
4. Fill the column with the chosen stationary phase (upper phase).
5. Pump the mobile phase (lower phase) into the "head" end of the inlet column at a flow rate of 1.0 mL/min and a rotation of 126 × *g*. The column is filled if the mobile phase is eluting at the tail outlet.
6. Inject the sample solution from the sample loop into the column through the sample port.
7. Operate the system at a flow rate of 1 mL/min and a rotation of 1000 rpm.
8. Collect the eluate into test tubes at a fraction size of 10 mL using a fraction collector.
9. Analyze fractions either by thin layer chromatography as described in Subheading 3.2.5 or by HPLC–DAD–MS/MS (*see* Subheading 3.2.6). Combine fractions with same content, transfer them to preweight glasses, and evaporate them to dryness using the SpeedVac Plus.

#### 3.2.5 Fraction Analysis by Thin Layer Chromatography

1. Perform thin layer chromatography on TLC aluminum plates (Silica gel $60F_{254}$) using ethyl acetate/acetone/water/formic acid (60:40:2:1) as mobile phase.
2. Dip dried TLC plates into a 1% ethanolic iodine solution and dry it using the hairdryer. Destruxins appear as brownish spots against a yellow-brown background (*see* **Note 28**).

#### 3.2.6 Fraction Analysis by HPLC–DAD– MS/MS

1. Perform the separation of analytes on a Zorbax SB-C18 column equipped with a C18 cartridge used as guard column.
2. Run the analysis using a gradient method with the solvents water (A) and acetonitrile (B) and a gradient of $t=0$ min 70% A; $t=4$ min 2% A; $t=5.3$ min 2% A, $t=5.4$ min 70% A. Between runs the column is equilibrated with 70% A for 4 min. The flow rate is set at 0.3 mL/min and the oven temperature at 23 °C.
3. Set the injection volume at 2 μL and record the chromatograms at 210 nm.
4. Perform MS experiments in positive ESI-mode using following experimental parameters: spray voltage 4500 V; nebulizer gas ($N_2$) flow set to 3.5 bar; capillary exit voltage 116.9 V; dry gas ($N_2$) flow 10 L/min with a temperature of 300 °C. A spectral scan range 100–1000 $m/z$ with a maximum accumulation time of 50 ms is applied.

#### 3.2.7 Sample Preparation from Culture Filtrate

1. Precondition SPE cartridges with 1 mL of methanol followed by 1 mL of distilled water by centrifugation for 1 min at 247 ×$g$.
2. Apply a 1 mL aliquot of the culture filtrate sample and wash cartridges with 1 mL of 40% methanol (v/v). Perform both steps by centrifugation for 1 min at 247 ×$g$.
3. Elute destruxin analytes with 1 mL of 85% methanol (v/v) by centrifugation for 2 min at 102 ×$g$.
4. Transfer the obtained eluate to HPLC vials.

#### 3.2.8 UHPLC–DAD Method for Destruxin Quantification from Culture Filtrate

1. Perform the separation of analytes on a Zorbax Eclipse XDB-C18 rapid resolution column, guarded with an inline filter.
2. Run the analysis with a flow rate of 0.3 mL/min using a gradient method: $t=0.0$ min, 75% A; is $t=3.5$ min, 60% A; is $t=4.5$ min, 50% A; is $t=5.5$ min; 35% A, is $t=6.5$ min, 5% A; is $t=7.0$ min, 2% A; is $t=12.0$ min, 2% A.
3. Reequilibrate the column between single runs for 12 min using starting conditions.
4. Set the injection volume at 2.5 μL and record the chromatograms at 210 nm.
5. Run all samples with these HPLC conditions (*see* Fig. 5).

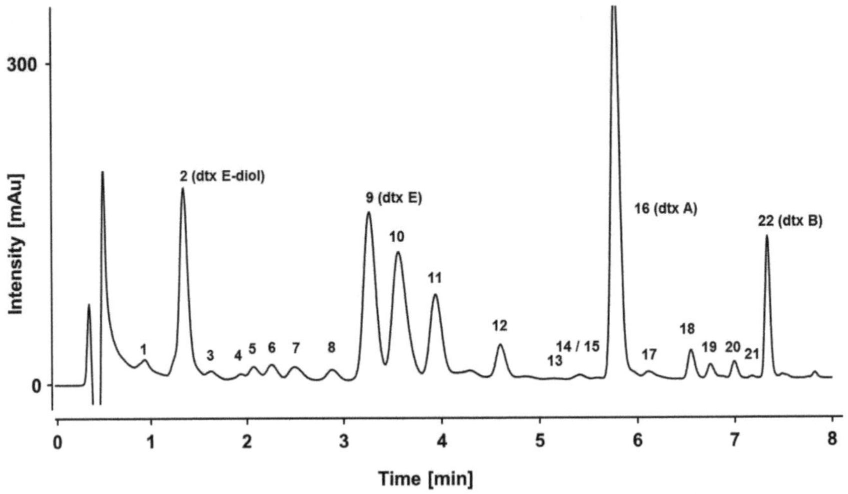


**Fig. 5** Representative UHPLC–DAD chromatogram of a fully developed *M. brunneum* BIPESCO 5 culture broth sample under optimized chromatographic conditions. Twenty-two chromatographic peaks were separated within little more than 7 min. Reproduced from Taibon et al. [13] with permission from Springer Science and Business Media

#### *3.2.9 UHPLC–DAD–QTOF–MS/MS Method for Destruxin Identification from Culture Filtrate*

1. Use for MS/MS experiments a Bruker micrOTOF-QII mass spectrometer.
2. Perform experiments in positive ESI mode with following parameters: capillary energy 4500 V, nebulizer gas 23.2 psi, dry gas 6.0 L/min at a temperature of 200 °C, scan range 50–1500 *m/z* with a scan rate of 2 Hz. Fragmentation is performed in automatic mode with a collision energy of 10–25 V.
3. If available use for destruxin identification reference material. Further dtx congeners can be tentatively assigned by analyzing TOF–MS data using exact masses and specific fragmentation pattern in comparison with literature data (*see* refs. 4, 5, 13) (Table 1).

#### *3.2.10 Assay Validation*

1. Use methanolic spike solutions to prepare final calibration levels by 1:10 dilution in either methanol or culture medium. Prepare matrix samples as described in Subheading 3.2.7, methanolic standards were used without further purification. Measure all levels three times to obtain calibration curves.
2. Prepare spiked samples for recovery experiments by SPE and measure all samples in triplicates on 3 different days.
3. Assess from this experimental setup the validation parameters LOD, LOQ, precision, accuracy, intra- and interday repeatability of the method.

**Table 1**
**UHPLC–DAD–TOF–MS/MS-based characterization of destruxin (dtx) congeners in a fully developed *Metarhizium brunneum* BIPESCO 5 culture broth sample**

| Peak | Retention time (min) | Masses in positive ion mode $[MH]^+$ | | Molecular formula | Destruxin derivative series | Destruxin derivative assignment[b] |
|---|---|---|---|---|---|---|
| | | Parent ion | Fragment ions[a] | | | |
| 2 | 1.34 | 612.3638 | 499, 471, 386 | $C_{29}H_{49}N_5O_9$ | X | E-diol[RS] |
| 4 | 1.94 | 596.3644 | 455, 384, 356 | $C_{29}H_{49}N_5O_8$ | desMe-X | Desmethyl C[LI] |
| 5 | 2.07 | 580.3334 | 467, 439, 368 | $C_{28}H_{45}N_5O_8$ | $X_2$ | $E_2$[LI] |
| 6 | 2.26 | 610.3446 | 497, 469, 370 | $C_{29}H_{47}N_5O_9$ | $X_2$ | $D_2$[LI] |
| 7 | 2.50 | 626.3745 | 513, 485, 400 | $C_{30}H_{51}N_5O_9$ | $X_1$ | $Ed_1$[LI] |
| 8 | 2.88 | 624.3625 | 511, 483, 370 | $C_{30}H_{49}N_5O_9$ | X | D[LI] |
| 9 | 3.26 | 594.3525 | 481, 453, 340 | $C_{29}H_{47}N_5O_8$ | X | E[RS] |
| 10 | 3.56 | 624.3653 | 511, 483, 370 | $C_{30}H_{49}N_5O_9$ | X | d-isomer[TE] |
| 11 | 3.94 | 630.3294 | 517, 489, 376 | $C_{29}H_{48}ClN_5O_8$ | X | Cl[LI] |
| 12 | 4.59 | 564.3392 | 451, 423, 324 | $C_{28}H_{45}N_5O_7$ | $X_2$ | $A_2$[TE] |
| 13 | 5.14 | 638.3787 | 525, 497, 384 | $C_{31}H_{51}N_5O_9$ | $X_1$ | $D_1$[LI] |
| 14 | 5.41 | 564.3395 | 451, 423, 352 | $C_{28}H_{45}N_5O_7$ | $X_2$ | $A_2$[TE] |
| 15 | 5.59 | 566.3545 | 467, 439, 340 | $C_{28}H_{47}N_5O_7$ | | $A_3$[TE] |
| 16 | 5.77 | 578.3559 | 465, 437, 352 | $C_{29}H_{47}N_5O_7$ | X | A[RS] |
| 17 | 6.12 | 566.3545 | 467, 439, 340 | $C_{28}H_{47}N_5O_7$ | | $A_3$[TE] |
| 18 | 6.55 | 580.3718 | 481, 453, 368 | $C_{29}H_{49}N_5O_7$ | desMe-X | Desmethyl B[LI] |
| 19 | 6.75 | 580.3696 | 552, 467, 439 | $C_{29}H_{49}N_5O_7$ | X | Dihydro A[LI] |
| 20 | 7.00 | 580.3719 | 467, 439, 368 | $C_{29}H_{49}N_5O_7$ | $X_2$ | $B_2$[LI] |
| 21 | 7.18 | 592.3692 | 479, 451, 366 | $C_{30}H_{49}N_5O_7$ | $X_1$ | $A_1$[LI] |
| 22 | 7.34 | 594.3837 | 481, 453, 340 | $C_{30}H_{51}N_5O_7$ | X | B[RS] |

Reproduced and modified from Taibon et al. [13] with permission from Springer Science and Business Media
Molecular formulas are derived from high resolution TOF MS data; tentative dtx congener assignment is following both published retention time and fragmentation pattern rules (*see* refs. 4, 5). Chromatographic peaks 1 and 3 are not included in the listing because they show no dtx-like fragmentation pattern

[a]For the sake of clarity, fragment ions are given in unit masses

[b]Assignment state (as superscript label): RS reference standard comparison, LI tentative assignment with the aid of literature data comparison (*see* refs. 4, 5), TE tentative assignment without the aid of literature data comparison

## 4 Notes

1. Adjust pH with either sodium hydroxide or phosphoric acid if needed.
2. Diluted oosporein solutions in methanol (<5 mg/L) are not stable within 24 h and a loss of the analyte is observed, which can be explained by complex formation with both sample matrix and glass surface. The oosporein anion can be only stabilized in solution by complex formation with the Britton–Robinson buffer system.
3. Vivaspin concentrators are disposable ultrafiltration devices for the concentration of biological samples. The used membrane consists of an acetylated cellulose (CTA) material with a cutoff of $M_r$ 10,000.
4. 2-iodobenzoic acid is chosen as IS because it has a p$K_a$ value close to that of oosporein and it has a well detectable UV response at 287 nm. Furthermore, it is commercially available at a high purity.
5. The mobile phase should be prepared fresh every day. After development, the plate has to be dried using a hairdryer to remove mobile phase residues such as formic acid, which could interfere with the derivatization reagent.
6. The selection of the appropriate two-phase solvent system is done based on the partition coefficients for dtx A, B, and E ($K_A$, $K_B$, and $K_E$) and the average partition coefficient $K_{ABE}$. Using the lower phase as mobile phase and the upper phase as stationary phase the elution mode for the HSCCC separation is "head to tail." Operating the HSCCC in this mode optimal $K$ values should range between 0.5 and 1.0. Although $K$ values for the system used are not all in this range, this system gives satisfactory separations.
7. An additional calibration function of the matrix sample (culture filtrate) is necessary to exclude a matrix effect.
8. Filtering small quantities of conidial suspensions to remove mycelial fragments (*see* ref. 16).
9. The pH value of the liquid medium II should be pH 6.
10. The final purification step over filtration gauze ensures that the supernatant is free of fragments of the mycelium which were not retained from the cotton cloth.
11. Mixing the bags after 5 days incubation ensures a homogeneous colonization of the kernels.
12. Changes in the volume are carefully monitored and are found to be negligible.

13. Changes in the volume are carefully monitored and are found to be negligible. Repeated sonication is done to ensure a quantitative extraction of the analytes.
14. The addition of an internal standard to the extraction solvent allows monitoring of the whole extraction procedure.
15. Samples extracted at room temperature remain stable for at least 7 days.
16. The use of an acidified mobile phase system allows a successful oosporein analysis. The combination of acetic acid and formic acid (1:9 ratio v/v) added to both mobile phase constituents (1 %) results in a pH of the water phase of $2.13 \pm 0.05$. Elevated pH values show a decline in the symmetry of the analyte peak. Thus, keeping the mobile phase $pH < 2.5$ is a prerequisite for maintaining a reasonable peak shape of oosporein. Those additives are preferred to other modifiers (e.g., trifluoroacetic acid) because their use still allows hyphenation to mass spectrometry (HPLC–DAD/MS).
17. Although sample matrices present a multitude of chromatographic peaks, the peaks of oosporein and IS are well separated from other constituents. Even after numerous analyses of samples no loss of the column performance is observed.
18. All spiking levels used should be adjusted to the oosporein concentration found in the respective matrix.
19. The pH value of the liquid medium II should be pH 6.
20. The use of organic solvent and culture filtrate at a ratio of 3:1 is necessary to achieve a complete phase separation during extraction. To guarantee the complete extraction of destruxins from culture broth each aliquot should be extracted five times.
21. The wash step of the organic layers with water ensures the removal of unwanted polar substances.
22. Water-free sodium sulfate is used as desiccant to remove water from organic solvents.
23. By recycling of the dichloromethane the total solvent consumption can be kept below 5 L dichloromethane/10 L culture filtrate.
24. Transfer the Sephadex LH-20 material into a 1000 mL Erlenmeyer and use as much solvent as needed to cover the material.
25. Prior to application to the column the dissolved crude extract should be filtered through cotton batting.
26. Both methods can be used for fraction analysis. Thin layer chromatography offers the possibility of a multiple analysis of fractions in a minimum of time and allows easily pooling fractions containing same analytes. HPLC–DAD–MS/MS is an

alternative method, which enables further destruxin identification and purity control.

27. If the upper phase of the solvent system is used as stationary phase the operation mode is "head to tail," if the upper phase is used as the mobile phase the operation mode is "tail to head."
28. TLC plates can also be sprayed manually using a rubber pump. By dipping the plate a more homogeneous result can be achieved.

## Acknowledgements

This research has been supported by the European Community's Seventh Framework Programme grant (FP7_ENV.2011.3.1.9-1 ECO-INNOVATION, INBIOSOIL, Grant Agreement No. 282767). We grateful acknowledge the contribution from PD Dr. Christoph Seger, Dr. Sonja Sturm and Prof. Dr. Hermann Stuppner (all University Innsbruck) who provided their expertise within this project.

### References

1. Seger C, Sturm S, Längle T, Wimmer W, Stuppner H, Strasser H (2005) Development of a sensitive high-performance liquid chromatography-diode array detection assay for the detection and quantification of the *Beauveria* metabolite oosporein from submerged culture broth and bio-control formulations. J Agric Food Chem 53:1364–1369
2. Seger C, Längle T, Pernfuss B, Stuppner H, Strasser H (2005) High-performance liquid chromatography-diode array detection assay for the detection and quantification of the *Beauveria* metabolite oosporein from potato tubers. J Chromatogr A 1092:254–257
3. Seger C, Eberhart K, Sturm S, Strasser H, Stuppner H (2006) Apolar chromatography on Sephadex LH-20 combined with high-speed counter-current chromatography. High yield strategy for structurally closely related analytes-Destruxin derivatives from Metarhizium anisopliae as a case study. J Chromatogr A 1117:67–73
4. Seger C, Sturm S, Stuppner H, Butt TM, Strasser H (2004) Combination of a new sample preparation strategy with an accelerated high-performance liquid chromatography assay with photodiode array and mass spectrometric detection for the determination of destruxins from *Metarhizium anisopliae* culture broth. J Chromatogr A 1061:35–43
5. Jegorov A, Havlícek V, Sedmera P (1998) Rapid screening of destruxins by liquid chromatography/mass spectrometry. J Mass Spectrom 33:274–280
6. Kershaw MJ, Moorhouse ER, Bateman R (1999) The role of destruxins in the pathogenicity of *Metarhizium anisopliae* for three species of insect. J Invertebr Pathol 74:213–223
7. Loutelier C, Cherton JC, Lange C, Traris M, Vey A (1996) Studies on the dynamics of the production of destruxins by *Metarhizium anisopliae* direct high-performance liquid chromatographic and fast atom bombardment mass spectrometric analysis correlated with biological activity tests. J Chromatogr A 738:181–189
8. Potterat O, Wagner K, Haag H (2000) Liquid chromatography–electrospray time-of-flight mass spectrometry for on-line accurate mass determination and identification of cyclodepsipeptides in a crude extract of the fungus *Metarrhizium anisopliae*. J Chromatogr A 872:85–90
9. Wang C, Skrobek A, Butt TM (2003) Concurrence of losing a chromosome and the ability to produce destruxins in a mutant of *Metarhizium anisopliae*. FEMS Microbiol Lett 226:373–378
10. Wang C, Skrobek A, Butt TM (2004) Investigations on the destruxin production of the entomopathogenic fungus *Metarhizium anisopliae*. J Invertebr Pathol 85:168–174

11. Morais RP, Lira SP, Seleghim MHR, Berlinck RGS (2010) A method for dextruxin analysis by HPLC-PDA-ELSD-MS. J Braz Chem Soc 21:2262–2271
12. Hsiao Y, Ko J (2001) Determination of destruxins, cyclic peptide toxins, produced by different strains of *Metarhizium anisopliae* and their mutants induced by ethyl methane sulfonate and ultraviolet using HPLC method. Toxicon 39:837–841
13. Taibon J, Sturm S, Seger C, Parth M, Strasser H, Stuppner H (2014) Development of a fast and selective UHPLC-DAD-QTOF-MS/MS method for the qualitative and quantitative assessment of destruxin profiles. Anal Bioanal Chem 406: 7623–7632
14. Liu C, Huang S, Tzeng Y (2004) Analysis of destruxins produced from Metarhizium anisopliae by capillary electrophoresis. J Chromatogr Sci 42:140–144
15. Taibon J, Sturm S, Seger C, Werth M, Strasser H, Stuppner H (2015) Supercritical fluid chromatography as an alternative tool for the qualitative and quantitative analysis of *Metarhizium brunneum* metabolites from culture broth. Planta Med. doi:10.1055/s-0035-1557823
16. Newmeyer D (1990) Filtering small quantities of conidial suspensions to remove mycelial fragments. Fungal Genet Newslett 37:27

# Chapter 16

# Development of Biopesticides and Future Opportunities


## Travis R. Glare, Roma L. Gwynn, and Maria E. Moran-Diez



## Abstract

Biopesticides, pesticides based on living organisms or their extracts, are increasing in sales around the world, as synthetic pesticides are less available and environmental and health issues drive new approaches. Despite the increasing sales and use, there are still limitations that restrict more widespread uptake, such as slow to kill, cost, difficulties of production, lack of appropriate formulations, and reputation based on previous poor performance of biopesticides. Regulation continues to be problematic in many countries, as the processes are designed for evaluating chemistry rather than live organisms. Biopesticides do have a bright future, given the amount of investment currently in the area, improving products and growing need.



**Key words** Entomopathogens, Biopesticides, Regulation, Product development


## 1 Introduction

A number of recent reviews (e.g., [1–4]) have outlined the bright predictions for biopesticides sales in the coming years. Independent assessments have suggested increases of up to 15 % per annum worldwide, although any such figures are hard to verify. But these reports do indicate biopesticides may be entering a new era of mainstream use, rather than niche market products. This book has provided chapters addressing some of the technical requirements of biopesticide development, such as production, formulation, bioassay, and application. However, the process of getting products into markets is based on far more than just technical development. The commercialization process for taking effective microbial agents through to available biopesticides has many legislative and marketing issues, as well as some technical limits common to many potential products.

### 1.1 The Need

There is no doubt the need for new pest, weed, and disease control products is growing. Pests have been estimated to cause between 27 and 42 % losses in production for major crops around the world. This would rise to an estimated 48–83 % without crop protection


Travis R. Glare and Maria E. Moran-Diez (eds.), *Microbial-Based Biopesticides: Methods and Protocols*, Methods in Molecular Biology, vol. 1477, DOI 10.1007/978-1-4939-6367-6_16, © Springer Science+Business Media New York 2016

products, such as synthetic chemicals [5, 6]. The green revolution, the massive increase in food production in the past 50 years, is partly based on great increases in the use of synthetic pesticides (15–20 times) [5, 6]. It is likely that changes to climate patterns will exacerbate crop damage by pests and disease. The need for pesticides is increasing at a time when new synthetic agent leads are decreasing [2]. There is also pressure on current control approaches in terms of environmental and mammalian safety, with many countries moving to banning outright some groups of chemicals used in pesticides [1, 7].

### 1.2 Increasing Use of Microbial-Based Biopesticides

There are an increasing range of biopesticides being registered around the wrold (Table 1). It has been noticeable in the last 5 years that the major pesticide companies around the world have been acquiring small- to medium-sized biopesticides companies and/or products [8]. The acquisition of Agraquest and Prophyta by Bayer CropScience, Becker Underwood by BASF, Pace International by Valent BioSciences, and Pasteuria BioScience by Syngenta [8] shows the value of the growing market. At least one of these purchases was for over $1B US, although the company bought had more than just biopesticides as existing products. This reflects both the growing need for novel products as pesticides are withdrawn, and the recognition that the market is ready for these products. This has led to changes in the type of biopesticides that are likely to be successful. The largest pesticides companies are working in the largest agricultural and forestry markets, therefore most interest is shown in biopesticides with application for high value crops, such as horticultural crops with pests and disease including thrips, whitefly, powdery mildew, and *Botrytis*. This differs from the history of microbial-based biopesticides, which were more often niche, regionally developed products often targeting single pest species on minor crops.

Reasons for increasing use by growers and farmers include the following benefits:

- Efficacy against the target pest. Strain selection and new formulation and application techniques have increased the efficacy of many microbial-based products.
- Production efficacy. The yield and quality of the active agents has been improved in many cases.
- Host specificity. Host biological agents are more limited in host range than synthetic pesticides, which have made them more attractive for both environmental safety and registration viewpoints.
- Can be used in Integrated Pest Management (IPM) programs.

- Useful for resistance management. Resistance development to agents with an infective action (as opposed to those reliant on a toxin) has rarely been found following years of field application.
- Useful for residue management. In most cases, biological agents are not considered residues on produce.
- Growers—crop management, many biological agents have no worker reentry interval so growers can harvest when it is best for the crop.
- Worker safety. Microbial agents are screened for mammalian toxicity and not used if there are any issues.
- Favorable environmental footprint. In an era of concern over environmental pollution of all types, the biological and biodegradable nature of microbial-based pesticides is highly favorable.
- Use in organic production. Depending on the formulation ingredients biopesticides are suitable for use in certified organic production.

## 2 Regulation

Most countries have a regulatory process for pesticides. As has been well covered elsewhere [1, 7, 9, 10], many of these regulatory processes have not been updated for specific needs of microbial-based biopesticides, meaning some of the requirements are not fit for purpose.

Regulation is needed for any pesticide. Reasons include:

- Protection of the natural environment
- Protection of human safety
- Maintain consumer standards
- Protect farmers and growers by having quality standards

**Table 1**
**Examples of products from the USA**

| | Bioinsecticide | Biofungicide | Bioherbicide | Bionematicide | Other |
|---|---|---|---|---|---|
| Microorganism Bt[a] | 44 | – | – | – | – |
| Microorganism non-Bt | 18 | 41 | 5 | 3 | 1 |
| Botanical | 8 | 6 | 1 | 2 | 29 |
| Semiochemical | 56 | – | – | – | – |
| Other | 25 | 8 | 3 | 0 | 29 |
| Total | 151 | 55 | 9 | 5 | 59 |

[a] *Bt Bacillus thuringiensis*

- Protection of technological invention
- Protection of rights
- Maintain product standards

The issues of harmonization of regulations across the world continue to plague biopesticide development. Most jurisdictions have not developed specific guidelines for assessment of biopesticides based on live organisms, resulting in inappropriate registration procedures largely derived from assessing synthetic chemical-based pesticides. Some countries have developed specific guidelines, such as the USA Environmental Protection Agency (EPA).

Another complicating factor is that it can be difficult to import products based on live organisms into new regions, due to concerns around exotic organism introductions.

### 2.1 Data Requirements for Registration

For registration of biopesticides, each jurisdiction has requirements for the data package submitted. The EU and EPA requirements, for example, do currently differ, but have enough similarities that it is possible to generalize.

Data requirements for an active substance usually include:

1. Identity and purity
2. Physical and chemical or biological properties
3. Further information on use, production processes, and related areas
4. Analytical methods used to identify the active(s)
5. Human health effects
6. Residues (often confused with persistence)
7. Fate and behavior in the environment
8. Effects on nontargets
9. Summary of all

Data requirements for the formulated product:

1. Identity and composition of the formulation
2. Physical and chemical properties
3. Application, labeling, and packaging
4. Further information
5. Analytical methods
6. Efficacy data
7. Toxicology and exposure
8. Residues
9. Fate and behavior in the environment
10. Effects on nontarget organisms
11. Summary

The data on these requirements is compiled into a "dossier." Components of the dossier are used in risk assessment; hazard or exposure information about the active substance and/or product. It is generally required or at least good practice to have studies conducted in laboratories of GLP standard. The dossier will contain data from studies and trials, published papers which contain findings relevant to answering the regulatory question—either single papers or the 'weight of evidence,' specific pieces of information (e.g., "The product will be applied to cereals") and waivers, also known as scientific justifications, where it is explained that the data requirement is not relevant because of a specific reason—this use of waivers is essential for biopesticides.

## 3 Areas of Potential Improvement in Biopesticides

What would make biopesticides more effective or increase market share of effective biopesticides? Some of the identified limitations that have reduced biopesticide uptake include:

- Lack of highly virulent strains.
- Slow to kill.
- Environmental constraints.
- Lack of suitable stage for mass production or application.
- Complex life cycles of agents.
- Complex handling requirements.
- Variable effects, due to any combination of the above.
- Expensive in comparison to synthetic pesticides.
- High production and research costs.
- Lack of profits for companies.
- Regulatory constraints.
- Problems with formulations and marketing.
- Expectations are often of a chemical equivalent: fast acting, cheap, and broad spectrum.

### 3.1 Improving Regulation

As stated earlier, biopesticides are required to be registered in most markets. The regulations often used the same system as for chemical pesticides. In some cases, such as the EU, registration is a two-stage process, with both the active substance and the product registered separately. It can take 4–5 years to achieve registration and the cost of the full dataset for registration can be significant (over € 500M), although this is still cheaper than registration of synthetic chemical pesticides in most cases. There is a move around the world to harmonize biopesticide regulations but this is still in development in most countries.

Reflecting the farmers and growers interest in using biopesticides and the increased demand for these types of products, the FAO, a global organization, are updating and expanding their guidance for microorganisms, botanical and semiochemical-based pesticides and use of these technologies.

This new guidance document considers pest control agents based on microorganisms, botanicals, and semiochemicals. These are distinguished from conventional chemical pesticides by a combination of their active substance material and/or nature. The view that biopesticides have characteristics that require particular consideration for registration is shared by USA-EPA, the EU, and the OECD Biopesticide Steering Committee and many countries are involved in this work, developing a harmonized approach to 'biopesticide' registration (USA, Canada, EU, Japan, Australia, New Zealand). In acknowledgment that biopesticides are a special situation, specific biopesticides registration guidelines have also been developed by certain countries (Brazil, China, Ghana, Kenya, and Southeast Asia). However, in many countries, microorganisms, botanicals, and semiochemicals are evaluated and registered following the same system as for conventional chemical pesticides; this approach can pose an unnecessarily high regulatory burden to satisfy inappropriate testing requirements.

Harmonization of data requirements and of procedures for registration was recognized as an important step to facilitate the availability of microorganisms, botanicals, and semiochemicals. The guideline describes the basic data requirements and evaluation for field trial permit and registration for these technologies. This updated guidance will be available in 2016.

### 3.2 Strain Selection

There are a number of areas where advances may result in better biopesticide uptake. As detailed in some of the methods in this book, bioassay is a standard and necessary approach to biopesticide development. However, laboratory bioassay results do not always translate to field success, given the complexity of ecosystems and climatic effects. But efficient bioassay can be the crucial step in separating potentially useful strains from the vast array of microbial candidates.

Many researchers are now looking for methods to more rapidly identify the most appropriate strain of a microbial for use in biopesticides. Less than 1 % of candidate isolates eventually make successful products, so methods that can improve the search approach are sought. Recent approaches have included use of massive DNA sequencing to directly target activity-related genes, rather than testing each microbe. The success of such approaches is still to be seen.

### 3.3 Production and Formulation

Production remains one of the key areas for making biopesticides cost effective. Microbial agents are often very effective when applied at high rates, but the cost of production precludes their use. In this book, several production protocols are outlined.

Further improvements and efficiencies gained in production will continue to make biopesticides more successful.

Formulation has provided some of the more effective improvements in the biopesticides area in the last decades. The use of prills and emulsions, techniques covered in this book, continues to improve the application, persistence and efficacy of biopesticides. Seed coating, also covered herein, is a new and increasingly attractive method to deliver biopesticides, especially in the soil. Seed coating is increasingly attractive as more agents are shown to be rhizosphere colonizers or even capable of endophytic colonization.

### 3.4 Application and Monitoring

Following production and formulation, the microbes and their bioactives need to be delivered to the target pest. This is one of the most challenging steps in the use of biopesticides, partly because most application techniques were originally developed for synthetic pesticides, not live organisms. Ideally application establishes the active agents in contact with the pest and/or maintains activity for several weeks. Spray applicators have routinely been used for aboveground application. There is increasing focus on the basics of spray application with microbial agents, including specialized equipment, optimal droplet size, and targeted application. Application subsurface is more problematic, as delivery is difficult without damaging the soils and plants. However, once delivered subsurface, persistence is often higher than aboveground applications.

### 3.5 Quality Control

Product variability has been a major issue in biopesticide development, but quality control to standardized batches is now generally recognized and incorporated into production systems. Ideally, each production batch is tested for efficacy against a target insect, stability, and propagules (yield).

### 3.6 Environmental and Mammalian Safety

One of the driving forces behind the increasing sales of biopesticides has been market pull. Biological agents are perceived as more inherently safe than synthetic chemical pesticides. However, all products and agents still must pass rigorous safety testing for most regulatory regimes. This book provides some methods around safety evaluations. It is likely to become more of a focus as more is understood about the mode of action of biological agents, the increase in use of bioactive directly rather than whole organisms, and as part of the wider public perception and concerns over risks.

## 4 Innovative Approaches

While incremental improvements are constantly made across all areas of delivering effective biopesticides based on microbial agents, there are several approaches which offer new paradigms for using microbial agents. Aspects of these are covered in this book.

### 4.1 Endophytes

One of the rapidly growing areas of investigation is the exploitation of plant endophytes. Fungi and bacteria are commonly found within plants and can confer significant pest and disease resistance [11]. In New Zealand, the pastoral industry almost exclusively used grass with *Epichloë* spp. endophytes, which confer pest, disease, and drought tolerance [2].

Many microorganisms used in biopesticides also deliver a number of additional benefits beyond virulence to a primary target. For example *Trichoderma* spp. can enhance the uptake of soil macro- and micronutrients by plants and substantial plant growth benefits in the absence of a disease. Entomopathogenic fungi can also have antagonistic activity against plant pathogens attacking the same crop. Endophytes are all about chemistry. They produce a range of bioactive secondary metabolites (such as alkaloids). The type of alkaloids produced depends on the strain of fungus present. The host plant has a major effect on the quantity of alkaloids. The chemistry of endophytes is diverse and complex.

The literature on endophytes is growing exponentially currently, suggesting new products or plant varieties are likely to emerge.

### 4.2 Bioactives

The use of just the active component of biocontrol microbial agents has long been attractive. The most successful microbial control agent in used commercial products, *Bacillus thuringiensis*, kills insects through toxic proteins, rather than an infective action. Serenade, a products based on *Bacillus subtilis*, contains live microorganisms and a combination of known and novel lipopeptides (agrastatins). In these cases, it may not always be necessary that the microbe is alive for a product to be effective, as the bioactive effect is present due to secondary compounds. Microbial secondary compounds can be produced and optimized in fermentation, which can make the process very amendable to scale-up and optimization. Microbial secondary compounds can also have more of a synthetic pesticide equivalence, making them easier to incorporate in current pest management practice. Depending on the nature of the secondary compounds, there may need to be consideration of residues on food and potential of resistance development in the targets.

## 5 Integrated Pest Management (IPM)

Biopesticides fit IPM systems well, usually being compatible with other biologically based controls (e.g., parasitoids/predators). Integrated pest management is not a new idea, but its application is dependent on having a range of tools that can be combined to reduce pest impacts below economic thresholds. These tools can include environmental safe chemicals, semiochemicals, plant varieties, physical methods, decision support tools including monitoring and biopesticides. The main ingredient of IPM is that the activities and tools act together to lead to pest management. The European Union has enacted legislation designed to strongly encourage the use of IPM [1].

Several companies are actively promoting the combined use of biopesticide and synthetic pesticide, such as the company Bayer with Votivo, based on *Bacillus firmus* for nematode control, combined with a synthetic insecticide, Poncho, as a seed treatment.

## 6 Summary and Future Directions

As demonstrated by the increasing sales, acquisition of small production companies by large companies, and the new products entering the market, the future for biopesticides looks very promising. Largely driven by market need, with many current synthetic chemical pesticides used in control being withdrawn, biopesticides have become the main pesticides used in some sectors in some regions. However, as detailed in Glare et al. [2], there are specific areas where research can lead to step change in the uptake of biopesticides. The review recommended:

- More research emphasis on delivery and persistence of biopesticides in the field. The aspirational target for persistence on foliage was put at 21 days and, in soil, persistence at the site of pest occurrence, rather than just persistence.
- More research emphasis on the chemistry of bioactives from microorganisms. This was seen as an area underdeveloped.
- More strategic selection of target pests and markets. The economics of biopesticide use can still be constraining so targeting of high value markets and highly susceptible pests is necessary.
- Continued investment in expertise for the discovery, development, and implementation of biopesticides. Biopesticides remain an underresearched area and additional investment in research from fundamental to applied subjects will reap benefits.
- Registration and legislative changes to better align data requirements with the features of biopesticides.

Similarly, there are features of successful biopesticide development that are common. Some of these are as follows:

- Take-up of biocontrol agents often depends on commitment and drive of scientist involved. It is surprising how important a product champion can be to the success of microbial biopesticide development, especially in the prototype stage.
- Well-defined end user demand and market position.
- Products developed in partnerships with commercial producers. Prototype products are often developed by researchers in

public institutions, with a lack of commercial knowledge or developed pathways to market. The combination of commercial acumen and research capability is crucial.

- Market demand for products. This includes realistic assessment of competing products and costs.
- Policy framework encourages uptake. As discussed herein, regulation designed specifically to consider biopesticides can be an advantage and cost savings.
- A pragmatic approach by goverments to regulations and registration.
- Government funds work and/or subsidizes product. Development of biopesticides from the many possible agents is expensive and can fail. Public investment in research leads to more products being developed.
- Support from researcher/first developers. It has been demonstrated many times that involvement of researchers after the first stages of commercialization improves the success rate.
- Good quality control of final product. The quality of products that reach the user is more of an issue for those based on live organisms than other forms of pesticides.

Progress has clearly been made. There are many new products coming on to the market. Technological developments are continuing to overcome impediments. However, biopesticides have not yet reached their potential, even though all predictions suggest biopesticides will outperform other pest control options in terms of market share increases in the near future.

We see a bright future for biopesticides, if the research and industry groups can think bigger and act united, better communicate the positive messages about biopesticides, and demonstrate their ability to control pests effectively and economically.

## References


1. Chandler D, Bailey AS, Tatchell GM, Davidson G, Greaves J, Grant WP (2011) The development, regulation and use of biopesticides for integrated pest management. Philos Trans R Soc B 366:1987–1998
2. Glare TR, Caradus J, Gelernter WD, Jackson TA, Keyhani NO, Köhl J, Marrone PG, Morin L, Stewart A (2012) Have biopesticides come of age? Trends Biotechnol 30:250–258
3. Koul O (2011) Microbial biopesticides: opportunities and challenges. CAB Rev: Perspect Agric Vet Sci Nutr Nat Resour 6, No. 056
4. Nollet LML, Rathore HS (eds) (2015) Biopesticides handbook. CRC Press, London
5. Oerke EC, Dehne HW (2004) Safeguarding production—losses in major crops and the role of crop protection. Crop Prot 23:275–285
6. Oerke EC (2006) Crop losses to pests. J Agric Sci 144:31–43
7. Bailey KL, Boyetchko SM, Längle T (2010) Social and economic drivers shaping the future of biological control: a Canadian perspective on the factors affecting the development and use of microbial biopesticides. Biol Control 52:221–229
8. Beer A (2012) Company mergers, acquisitions and crop protection deals in 2012. Agrow Annu Rev Xi–Xiii

9. Kabaluk JT, Svircev AM, Goettel MS, Woo SG (eds) (2010) The use and regulation of microbial pesticides in representative jurisdictions worldwide. IOBC Global, Rome, Italy, 99 pp
10. Sundh I, Goettel MS (2013) Regulating biocontrol agents: a historical perspective and a critical examination comparing microbial and microbial agents. BioControl 58:575–593
11. Porras-Alfaro A, Bayman P (2011) Hidden fungi, emergent properties: endophytes and microbiomes. Annu Rev Phytopathol 49:291–315

ERRATUM TO

# Methods for the Evaluation of the Bioactivity and Biocontrol Potential of Species of *Trichoderma*

**Johanna Steyaert, Emily Hicks, Janaki Kandula, Diwakar Kandula, Hossein Alizadeh, Mark Braithwaite, Jessica Yardley, Artemio Mendoza-Mendoza, and Alison Stewart**

Erratum to: Chapter 3 in Travis R. Glare and Maria E. Moran-Diez (eds.), *Microbial-Based Biopesticides: Methods and Protocols*, Methods in Molecular Biology, vol. 1477, DOI 10.1007/978-1-4939-6367-6_3, © Springer Science+Business Media New York 2016

**Erratum DOI 10.1007/978-1-4939-6367-6_17**

Alison Stewart was not listed among the authors. The correct sequence of the author names is Johanna Steyaert, Emily Hicks, Janaki Kandula, Diwakar Kandula, Hossein Alizadeh, Mark Braithwaite, Jessica Yardley, Artemio Mendoza-Mendoza, and Alison Stewart

The updated original online version for this chapter can be found at DOI 10.1007/978-1-4939-6367-6_3

# Index



Travis R. Glare and Maria E. Moran-Diez (eds.), *Microbial-Based Biopesticides: Methods and Protocols*, Methods in Molecular Biology, vol. 1477, DOI 10.1007/978-1-4939-6367-6, 

MIX
Papier aus verantwortungsvollen Quellen
Paper from responsible sources
FSC® C105338

If you have any concerns about our products,
you can contact us on
**ProductSafety@springernature.com**

In case Publisher is established outside the EU,
the EU authorized representative is:
**Springer Nature Customer Service Center GmbH**
**Europaplatz 3, 69115 Heidelberg, Germany**

Printed by Libri Plureos GmbH
in Hamburg, Germany